Deserts and Desertification

Edited by Yajuan Zhu, Qinghong Luo and Yuguo Liu

Published in London, United Kingdom

IntechOpen

Supporting open minds since 2005

Deserts and Desertification
http://dx.doi.org/10.5772/intechopen.82931
Edited by Yajuan Zhu, Qinghong Luo and Yuguo Liu

Contributors
Yajuan Zhu, Fengmin Luo, Zhiming Xin, Junliang Gao, Yuan Ma, Huaiyuan Liu, Xing Li, Qinghong Luo, Qimin Chen, Miao He, Na Li, Raman Bala, Justine Nyaga, Sylus Musei, Abdi Zeila Dubow, Peter Hyde, Alex Mahalov, Guoqi Li, Shujun Li, Wenshan Shao, Yanyun Chen, Wang Yafang, Mehdi Rastgoo, Alireza Hasanfard, Yuguo Liu, Jinxing Zhou, Ming Cui, Jiufu Luo, Jinchao Feng, Wei Li, Linna Ma, Abiud L. Lucas Kaswamila, Fredy S. Mswima, Joseph Hitimana, Edward K. Mengich, Teresiah N. Kuria, Pauline Kimani

First published in London, United Kingdom, 2022 by IntechOpen
IntechOpen is the global imprint of INTECHOPEN LIMITED, registered in England and Wales, registration number: 11086078, 5 Princes Gate Court, London, SW7 2QJ, United Kingdom
Printed in Croatia

British Library Cataloguing-in-Publication Data
A catalogue record for this book is available from the British Library

Additional hard and PDF copies can be obtained from orders@intechopen.com

Deserts and Desertification
Edited by Yajuan Zhu, Qinghong Luo and Yuguo Liu
p. cm.
Print ISBN 978-1-83962-717-0
Online ISBN 978-1-83962-718-7
eBook (PDF) ISBN 978-1-83962-754-5

For EU product safety concerns:
IN TECH d.o.o., Prolaz Marije Krucifikse Kozulić 3, 51000 Rijeka, Croatia, info@intechopen.com or visit our website at intechopen.com.

Meet the editors

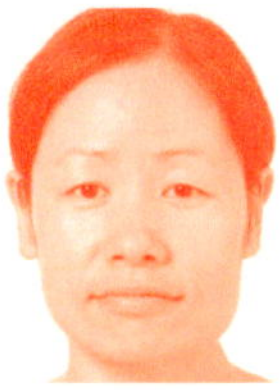

Yajaun Zhu received her Ph.D. from the Institute of Botany, Chinese Academy of Sciences in 2007. She is currently an associate professor at the Institute of Desertification Studies, Chinese Academy of Forestry. She has published sixty-four articles on desert ecology and land desertification control. She has published two books on desert plants and ecosystem management.

Qinghong Luo received her Ph.D. degree from Xinjiang Ecology and Geography Institute, Chinese Academy of Sciences in 2018. She is currently a professor at Xinjiang Academy of Forestry, China. She has published fifteen articles and three books on land desertification control.

Yuguo Liu received his Ph.D. from the Institute of Botany, Chinese Academy of Sciences in 2013. He is currently an associate professor at the Institute of Desertification Studies, Chinese Academy of Forestry. He has published twelve articles on restoration ecology and land desertification control.

Contents

Preface

Deserts are terrestrial ecosystems that cover more than one-fifth of Earth's land area. They are mainly distributed in subtropical and temperate zones with an arid climate. Sand dunes are typical landforms in deserts. Shelterbelt systems are built to fight desertification and defend against sandstorms. Desert vegetation consists of drought-resistant trees and shrubs. Algae crust plays an important role in stabilizing sand dunes, and shrubs have been widely used to fix moving sand dunes. Shrub sustainability is determined by their adaptation to the arid climate and factors such as water use strategy, stability, and soil seed bank. Moreover, some desert plants, such as jojoba, have important economic value.

Desertification is land degradation in arid, semi-arid, and sub-humid arid zones caused by climate change and human activities. Many measures have been taken to control land desertification. For example, eco-villages improve land productivity and strengthen communities' adaptation to climate change. The enhancement of woody floral diversity helps to restore desert vegetation. Engineering and biological measures decrease sand dune movement and maintain the security of railways.

This edited volume provides new insights into the understanding of deserts and desertification through a thoughtful mixture of viewpoints. Chapters include reviews, research, and case studies.

This book is a useful resource for professionals and researchers in ecology, forestry, land desertification control, and other related areas. We extend our sincere appreciation to the authors, who are from different countries, for their contributions. We also thank IntechOpen publishing for inviting us to be editors of this book. We would like to extend our special appreciation to Publishing Process Manager Ms. Marijana Josipovic for her superb support. We also thank Prof. Benli Liu for reviewing the first chapter.

Yajuan Zhu and Yuguo Liu
Institute of Desertification Studies,
Chinese Academy of Forestry,
Beijing, China

Qinghong Luo
Xinjiang Academy of Forestry,
Urumq, China

Section 1

Sand Dune and Sand Storm

Chapter 1

Bowing Sand, Dust, and Dunes, Then and Now–A North American Perspective

Peter Hyde and Alex Mahalov

Abstract

Dune fields of the present day, the Dust Bowl disaster of the 1930s U.S. Great Plains, and contemporary efforts to forecast, simulate, and understand dust storms have a striking, uniform commonality. What these apparently diverse phenomena have in common is that they all result from blowing sand and dust. This review paper unifies these three disparate but related phenomena. Its over-arching goal is to clearly explain these manifestations of windblown sand and dust. First, for contemporary dune fields, we offer reviews of two technical papers that explain the eolian formation and the continuing development of two major dune fields in southeastern California and northwestern Sonora, Mexico: the Algodones Dunes and the Gran Desierto de Altar. Second, historical, geological, meteorological, and socioeconomic aspects of the 1930s Great Plains Dust Bowl are discussed. Third, and last, we return to the present day to summarize two lengthy reports on dust storms and to review two technical papers that concern their forecasting and simulation. The intent of this review is to acquaint the interested reader with how eolian transport of sand and dust affects the formation of present-day dune fields, human agricultural enterprises, and efforts to better forecast and simulate dust storms. Implications: Blowing sand and dust have drastically affected the geological landscape and continue to shape the formation of dune fields today. Nearly a century ago the U.S. Great Plains suffered through the Dust Bowl, yet another consequence of blowing sand and dust brought on by drought and mismanagement of agricultural lands. Today, this phenomenon adversely affects landscapes, transportation, and human respiratory health. A more complete understanding of this phenomenon could (and has) led to more effective mitigation of dust sources, as well as to a more accurate predictive system by which the public can be forewarned.

Keywords: Dune fields of today, Dust Bowl of the 1930s Great Plains, science of dust storm formation, forecasting and simulating dust storms

1. Introduction

This review paper attempts to unify a single phenomenon – blowing sand and dust – as it concerns dune field formation, the 1930s Dust Bowl of the Great Plains, and the science of dust storm formation. This unification is first brought to life (in this paper's first section) by the presence of active sand dune fields, which, in effect, are repeating the same processes that led to sandstone formation, albeit in a nascent, formative manner. The second section of the paper continues with this theme of time as it portrays the 1930s Dust Bowl of the U.S. Great Plains. One of the papers summarized in this section relates the conditions of the Dust Bowl to a 1,000- year pageant of drought and moisture cycles in the western U.S. The third and last section of this review paper, grounded in the present day, first discusses two lengthy reports on dust storms, one of global extent with the other limited to Arizona. It then goes on to review contemporary efforts to understand, to monitor, and to forecast and simulate these dust storms. In sum, this review paper attempts to shed light on blowing sand and dust in prehistoric and historic time, and in the time of the present.

A cosmopolitan nuisance, blowing dust and sand affect virtually all semi-arid and arid landscapes [1], and has been doing so from deep geological time until the present day.

In their initial stage of formation, sand dunes depend on four related causes:

1. on the presence of a (usually) nearby source of sand, such as an oceanic beach;

2. on surface winds with sufficient speeds to suspend the sand and dust particles;

3. on these surface winds having a dominant, prevailing direction; and.

4. on the presence of a downwind receptor area capable of receiving and maintaining the transported sands [2–4].

The geological approach to blowing sand and dust, given an introductory summary above, can be more fully investigated by the curious reader through visiting these sand dune fields and through attending courses in the subject at the community college or at the university level. Furthermore, geological textbooks are widely available and can be studied independently [5–8]. Visiting present-day sand dunes offers the curious individual the advantages of travel and exploration throughout much of North America [9]: from northwestern Sonora, Mexico, to the Oregon coast, through the Midwest, and east as far as Cape Cod. This paper presents geological analyses of two such dune fields.

In addition to this approach, a second way to understand blowing dust and sand relies on historical reviews of particularly dusty periods. Although there are many to choose from, one of the better documented and the more instructive took place in the 1930s in the Great Plains of the U.S. and is known as the "Dust Bowl" [10]. This review paper presents some historical and meteorological insights into this nightmare.

Last, the paper explores modern-day dust storm magnitudes and frequencies, as well as their meteorological and landscape causes [11]. The paper goes further into this subject by describing how weather forecasters predict these storms and how atmospheric scientists come to understand their formation, transport, and eventual dissipation, [12, 13].

To summarize, this paper offers a three-fold synthesis of the natural phenomenon of blowing sand and dust. First, in the present day, how can extant dune fields shed light on their formation? Second, what can we learn about dust storms from an historical/scientific review of one of the worst recorded of such episodes, i.e. the infamous Dust Bowl in the Great Plains of the U.S. in the 1930s? Third what are the physical bases of dust storm formation and how do communities of meteorologists and atmospheric physicists study dust storms with the goal of reducing their deleterious effects on the land, on the atmosphere, on transportation, and on the respiratory health of the public?

2. North American sand dunes today

Many dune fields, but certainly not all, in North America, are found in semi-arid or arid regions. Because of this association, it is instructive to know about the four principal deserts of North America: namely, the Chihuahua, the Sonora, the Mojave, and the Great Basin deserts. Although each of the four is unique in its geographic distribution, in its landforms, in its local meteorology, and in its vegetation, all share a dry climate with sparse precipitation. **Figure 1**, which illustrates these deserts, shows that they are in northern and northwestern Mexico and the southwestern and western U.S.

Table 1 is a partial list of dunes in North America [11] that can be visited and explored today, all of which are protected as national parks, national recreation areas, or international biospheres. The table gives some basic geographical and meteorological information about each dune field: its area, its annual rainfall, and its low and high temperatures. Most of the areas given are for the extent of the active dune fields: exceptions are noted in the table. These dune fields are ordered from the southwest and the west coast, through the intermountain states, into the Midwest, and to the east coast of the U.S., with a far-northern outlier being the last in the table.

Geologists, geographers, and paleoclimate scientists have conducted many studies of these dunes, deducing the dynamics of their formation and of their present-day movements, paying close attention to the patterns of wind speeds, of wind directions, and of precipitation. Although this research has produced a considerable volume of journal articles and reports, this section will be limited to two

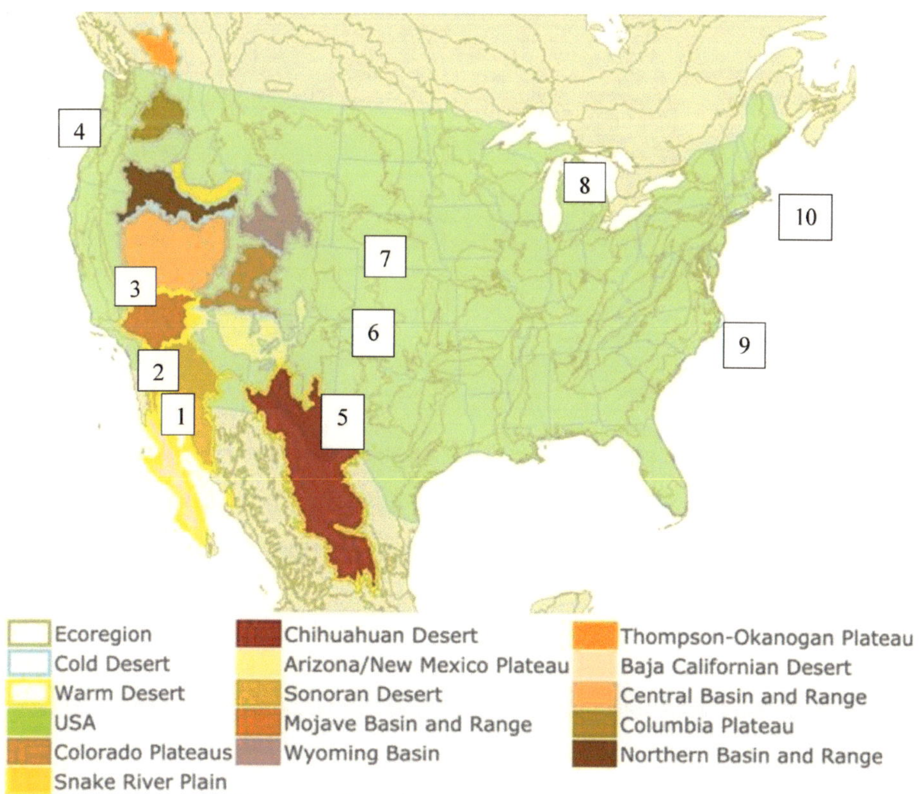

Figure 1.
Deserts of North America [11]: with 10 dune fields indicated.

#	Name	Location	Area (km^2)	Rainfall (mm)	Temperatures	Ref.
1	Gran Desierto de Altar	NW corner of Sonora, Mexico	5,7000	62	High: 45°C (113°F) Low: 10 °C (50°F)	[14]
2	Algodones Dunes	SE California, near Yuma, AZ	720	83	High: 42°C (107°F) Low: 11°C (43°F)	[15]
3	Death Valley National Park	Southeastern California	13,650 For the entire national park	38	High: 46°C (115°F) Low: 4°C (39°F)	[14]
4	Oregon Dunes	Oregon coast	134	1980	High: 22°C (71°F) Low: 0.5°C (33°F)	[16]
5	White Sands National Monument	Southcentral New Mexico	590	230	High: 36°C (97 oF) Low: 0°C (32 oF)	[14]
6	Great Sand Dunes National Park	Southcentral Colorado	603	283	High: 27°C (80°F) Low: < 0°C (32°F)	[14]
7	Nebraska Sand Hills	West-central Nebraska	51,000 For the entire region	430–580	High: 41°C (105°F) Low: −34°C (−30°F)	[14]
8	Sleeping Bear Dunes	Lower peninsula of Michigan	132	Rain: 726 Snow: 1970	High: 27°C (81°F) Low: 12°C (11°F)	[17]
9	Dunes of the Outer Banks	Coastal N.Carolina and Virginia	3,200 or the entire island chain	Rain: 1,245 Snow 15	High 31°C (88°F) Low: 3°C (38°F)	[18]
10	Dunes of Cape Cod	Coastal Massachusetts	34	Rain and Snow: 1,195	High: 26°C (78°F) Low: −4°C (25°F)	[19]
11	Kobuk Valley National Park	Alaska, north of the arctic circle	83 dune fields only	Rain and Snow: 331	High 19°C (66°F) Low: −24°C (−11°F)	[14]

Note: of these eleven dune fields, only the first three and the fifth are in a desert landscape.

Table 1.
Some dune fields in North America.

dune fields. These were chosen primarily because the authors are at least somewhat familiar with each one, given their proximity to Phoenix, Arizona. For example, in driving from Phoenix to San Diego on Interstate–8, one drives through one of these dune fields (Algodones). The corresponding author made this drive once when the sands were being suspended by turbulent winds such that the visibility was reduced to about 100 m. As for the other choice (Desierto de Altar), while the corresponding author has not been there, he has been very close while on a drive from Yuma, Arizona, through San Luis Rio Colorado (a city in north-eastern most Baja California, Mexico on the Arizona border), along the Colorado River delta south to the Gulf of California city of San Filipe. This drive affords a view of the Desierto de Gran Altar. Hence, the choice of these two nearby dune fields is a personal one.

2.1 Algodones dunes (southeastern California)

"Algodones" in Spanish means cotton. Fans of the Star Wars series of films may recognize the Algodones dune field—also known as the Imperial Dunes—as portions of the imaginary planet of Tatooine. This field is 72 km long by 10 km wide and extends along a northwest-southeast line that correlates with the prevailing northerly and westerly wind directions [20]. The weather is generally hot and dry, with the highest monthly average daytime temperature of 41.7°C (106.4°F), with monthly rainfall varying from 0.25 to 12 mm. Deserts can be cold, especially in winter nights, and these dunes are no exception, with the lowest monthly average temperature being 10.6°C (42.6°F).

The source of the sand of these dunes is the windblown beach sands of ancient Lake Cahuilla, itself formed by the meandering Colorado River as its waters periodically flowed into the Salton Sink. The most recent Lake Cahuilla covered much of the Imperial, Coachella, and Mexicali Valleys as late as 1450. The most popular theory holds that the Algodones Dunes were formed from windblown beach sands of Lake Cahuilla. The prevailing westerly and northwesterly winds carried the sand eastward from the old lake shore to their present location [20].

One study conducted in the 1980s, [21], thoroughly examined the formation and dynamics of these dunes, and a summary of their work follows. **Figure 1** from their paper, shown as **Figure 2**, gives the geographical setting of the area.

First, the authors use the term "draa", defined as a large sand dune hundreds of miles long and hundreds of feet high, often with smaller dunes that form on the leeward and windward faces. This term comes from the North African dialectal Arabic and Berber languages. Dunes are anything but static, and the migration rate of the Algodones Dunes has been measured at 0.09 meters per year. The dunes vary in width from 0.9 to 4.5 km, in length from 0.5 to 1.2 km, and in their inter-dunal distances from 0.7 to 1.5 km. Their journal article has many photographs and intricate diagrams of these dunes, but their work can be summarized by their conclusions.

1. The complex crescentic draas in the southern third of the Algodones dune field have crescentic and coalesced crescentic or star-like dunes superimposed on their stoss slopes. (Note that the term "stoss" means facing toward the direction from which an overriding glacier impinges, as in the stoss slope of a hill).

2. The draa is in equilibrium with the current wind regime in that: it is oriented perpendicular to the long-term primary wind resultant direction.

3. Although the draa is transverse to the average wind direction, winds approach the draa at an oblique angle a large percentage of the time. Because of this angle the draa exhibits features both of oblique and longitudinal bedforms as well as transverse bed forms.

4. The draa exhibits a complex bedform modified by a secondary airflow pattern.

5. The resultant migration direction of the draa is oblique to, and more easterly than, the resultant sand drift potential direction.

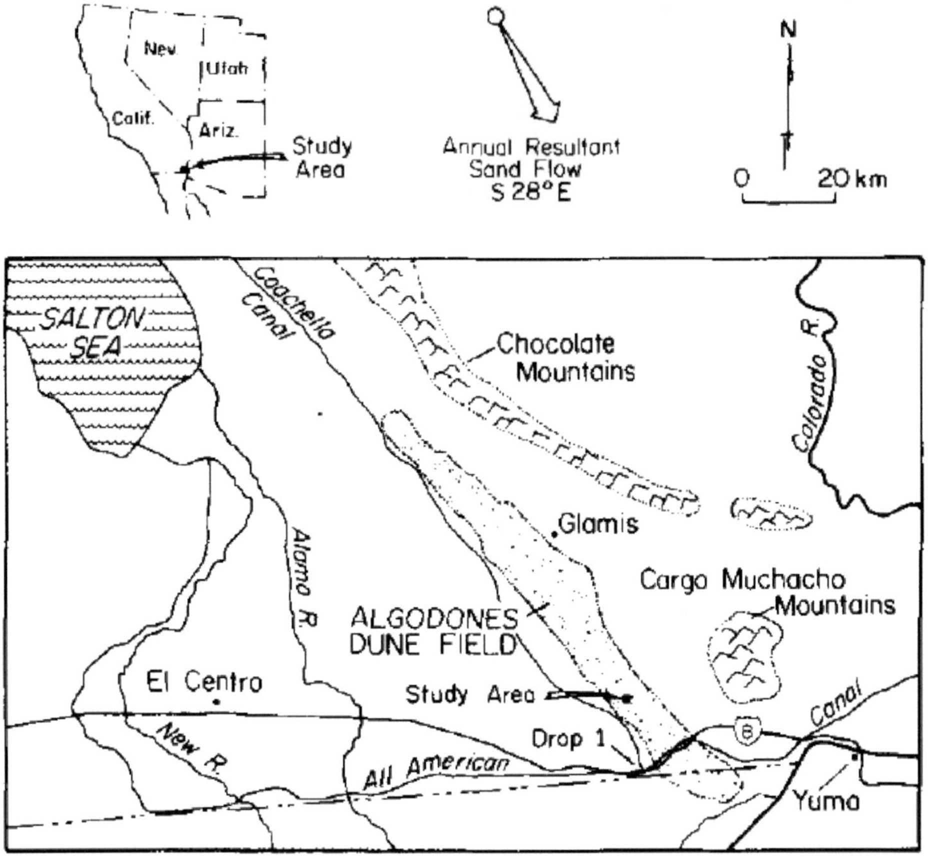

Figure 2.
Map of the Algodones dune field: The dashed line near the bottom is the border between the U.S. and Mexico; Figure 1 from [21].

6. The internal structure of the draa being generated at the base of the lee slope consists of two types: steeply dipping simple cross-strata, and compound crossstrata. Paleowind directions from such cross-strata should produce an internal structure that reflects both the direction of secondary airflow on the lee slope ... and the average wind direction that orients the draa (8). Draas can form both simple and compound crossstrata; their deposits will have great lateral variation.

In their work the authors thoroughly examine the interplay between wind speed and direction with the resultant dune configurations and movements. The paper employs geologically complex terminology and concepts and is probably unsuited for those readers without at least a moderate geological background. What the authors do not mention is that these dunes have served as a popular recreation area for nearby residents (e.g. Yuma, Arizona) to operate their dune buggies on weekends in winter.

Not far from the Algodunes Dunes is the Gran Desierto de Altar, an extremely dry and austere dune field in northwestern Sonora, Mexico that is a recognized biosphere preserve.

2.2 Gran Desierto de Altar

Translated as the "great desert of the altar (or shrine)", at the top of the Gulf of California, just across the Arizona border in Mexico, lies the Altar Desert, part of the El Pinacate y Gran Desierto de Altar, a biosphere reserve and world heritage site [22]. The desert is a small part of the much larger Sonoran Desert that encompasses much of the southwestern United States and northwestern Mexico. The Colorado River, which has its delta immediately to the west and just upwind of the Altar Desert, supplied abundant sand for the dunes' formation. This dune field is considered the largest and most active in North America. At one time centuries or millennia ago this dune field had as its northwestern most finger the Algodones Dunes, although today the two dune fields are separated by about 40 km. It includes the only active erg dune region in North America. (An "erg dune region" is a broad, flat area of desert covered with wind -swept sand with little or no vegetative cover.) This desert extends across much of the northern border of the Gulf of California, spanning more than 100 km east to west and over 50 km north to south. It constitutes the largest continuous wilderness area within the Sonoran Desert [22].

As with the Algodones Dunes, much research has been conducted in this immense sand sea, but a summary of only one investigation will be presented here [23], whose authors present a detailed map of the dunes (**Figure 3**) and the following conclusions.

Since middle Pleistocene time [roughly 1.3 mya], the Gran Desierto sand sea formed as an eolian deposit. Primary sand sources are the ancestral Colorado River flood plain and delta, the modern Colorado River flood plain, littoral sands from the Gulf of California, and local alluvial sources. Brief periods of eolian deposition (characterized by migration of crescentic dunes) were separated by long, stable intervals during which existing sand populations were modified from crescentic dunes to star and other complex dunes. In the present day, the low rates of sediment generation and transport in the Gran Desierto suggest that it is in a stable period, a situation that has probably existed during much of the late Holocene [the Holocene period is 12,000 years ago to the present].

The authors present numerous numerical analyses of the size and shapes of these dune fields, and they show many satellite images. All things considered, this is a highly technical paper that might elude the understanding of the casual reader. These are the only two technical summaries of geological work done in North

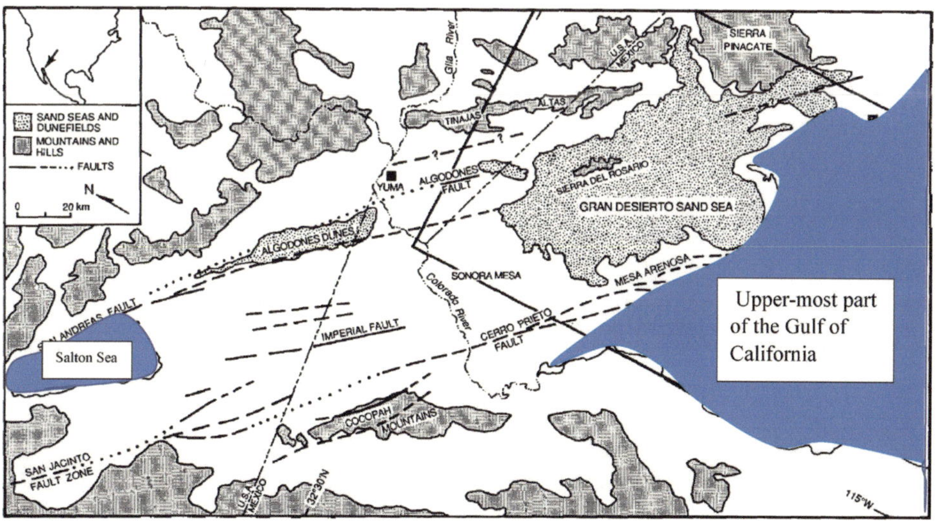

Figure 3.
Map and geological setting of the Gran Desierto de altar (Figure 1 of [23]): Note the proximity of the Algodones dunes, just 40 km northwest of the western edge of the Gran Desierto sand sea. Also notable is the U.S., Mexico border, the dashed line in the upper center of the map.

American dune fields that appear in this paper. The interested reader can find similar work on the others.

With the discussions of sandstone formation and extant dune fields completed, this paper moves on to review the historical and technical literature concerning the Dust Bowl of the 1930s.

3. The U.S. Great Plains dust bowl of the 1930s

Before examining the Dust Bowl, it is worth noting that on a global scale dust storms of major proportions have been documented at least for the last 200 years [14]. Furthermore, paleoclimate research has shed light on droughts and dust storms in the last several millennia [24]. **Table 2** gives eight of the more recent major dust storms [15]. Such storms in semi-arid and arid regions of the world have occurred throughout human history and long before, as the sandstone formations discussed in the first section of this paper attest. Of these eight dust storms, one took place in the Middle East, one in China, two in Australia, and four in the U.S. Immediately below the table is a brief damage assessment for each storm.

A. Black Sunday (Dust Bowl): 300 million tons of topsoil were lost.

B. Great Bakersfield dust storm: Swamp coolers were blown off the roofs of buildings. Windows were shattered. Trees, fences, and swamp coolers had blown down throughout the region. Below-grade freeways, canals, and creeks were buried in sand and dust. The storm resulted in five deaths and $40 million in damage. Over 25 million cubic feet of topsoil from grazing land alone was moved.

C. Melbourne dust storm: The winds brought down power lines and clogged electrical junction boxes with dust, causing them to short-circuit. Railroads could not function.

D. Interstate-5 dust storm: This date, the Friday after Thanksgiving, had heavier traffic than usual on Interstate-5. The dust storm caused a series of chain reaction accidents, which mainly occurred in five groups spread across 1.5 miles (2.4 km) of highway; while one 20-car pile-up occurred in the northbound lanes, the remainder of the crashes were in the southbound lanes. In total, 104 vehicles were involved in the accident, including 93 cars and 11 semi-trailer trucks. 17 people died in the accidents, and an additional 150 people were injured.

E. Australian dust storm: Vehicular and air transportation were disrupted. Ambulance services received around 140 calls from people having breathing

Name	Date	Affected regions	dam.
Black Sunday (Dust Bowl)	April 14, 1935	Texas, Oklahoma panhandles	A
Great Bakersfield dust storm	December 19–21, 1977	San Joaquin Valley, California	B
Melbourne dust storm	February 8, 1983	Victoria, Australia	C
Interstate-5 dust storm	November 29, 1991	San Joaquin Valley, California	D
Australian dust storm	September 23, 2009	S. Australia, New South Wales	E
China dust storms	Spring 2010	China, parts of southeast Asia	F
Arizona dust storm	July 5, 2011	South-central Arizona	G
Tehran dust storm	June 2, 2014	Tehran, Iran	H

Table 2.
Recent major dust storms: "dam." is for damages from the storms; see the letter keys below. Individual references are not shown here; merely typing the storm's name and date into a search engine brings up the information.

difficulties: more than 50 calls were made from Sydney, 50 from the state's west, 23 from the north and 12 from southern regions.

F. China dust storms: In the spring of 2010 many provinces of China were suffering from a severe drought that saw some 51 million citizens enduring water shortages. The series of dust storms that then ensued was in part a consequence of the desertification of extensive regions of the country. The annual direct economic losses attributable to desertification are estimated at US$ 7.7 billion. It is believed that the indirect economic losses arising from desertification amount to 43.5 billion US$ per year.

G. Arizona dust storm of 5 July 2011: Severe disruption of vehicular and air transportation, although no deaths or injuries were reported.

H. Tehran dust storm: 5 men were killed, more than 30 people were injured, and a few cars were destroyed. Falling trees and objects in balconies were destroyed. 65 electric lines of 20 KW were damaged and disconnected.

With these major dust storms enumerated, this review paper now proceeds to examine the Dust Bowl of the U.S. Great Plains, which started in 1930 and lasted for a decade. Severe drought hit the Midwest and southern Great Plains in 1930. Massive dust storms began in 1931. A series of drought years followed, further exacerbating the environmental disaster. By 1934, an estimated 35 million acres of formerly cultivated land had been rendered useless for farming, while another 125 million acres—an area roughly three-quarters the size of Texas—was rapidly losing its topsoil. Regular rainfall returned to the region by the end of 1939, bringing the Dust Bowl years to a close [15].

One of the affected states was Kansas, where in the 1940s an historian at the University of Kansas produced definitive studies of dust storms in the latter half of the nineteenth century [16–18]. Because the following discussion is limited to the Dust Bowl, these works have not been used in this paper, but they are included in the references for the interested reader. Instead, more contemporary research is relied on, and there is no shortage of such scholarship.

For instance, a team of Canadian researchers has assembled a comprehensive historical/scientific review of the Dust Bowl [19]. They present the geographical setting and the severity of the dustiest areas in their **Figure 1**, shown below as **Figure 4**. Although published seven years ago, their paper has perhaps the best and most comprehensive descriptions of the entire Dust Bowl saga, including its natural causes, its anthropogenic causes, and its disastrous consequences of soil erosion, of economic losses, and of forced mass migrations.

The underlying natural causes of the Dust Bowl have been succinctly described [19]:

Through data analysis and modeling, the authors state, that the causal mechanism for Dust Bowl era droughts on the Great Plains has been linked to ocean temperature anomalies. It appears that Pacific sea surface temperatures, especially as expressed by cold tropical temperatures during the La Niña phase of the El Niño Southern Oscillation, have the most direct influence.

The above-summarized work, a magnum opus, with copious geographical, historical, climatic, and economic analyses, is most suitable for the lay person. In sum, the authors cover much ground in a diversity of disciplines in a thorough, straightforward, and comprehensive manner.

The next summary, [24], though more localized, is of comparable analytic detail. Focusing on the region of northeastern Kansas and northwestern Missouri, a research team assembled rainfall records from 20 different cities and towns for the years 1850–2008 (**Figure 5**). They adjusted the records for 1850–1924 to account for negative biases in daily precipitation totals of less than 0.5 inches, which resulted in an overall increase of two percent above the historical records. In any case, the authors put the Dust Bowl into a broader historical perspective and conclude that

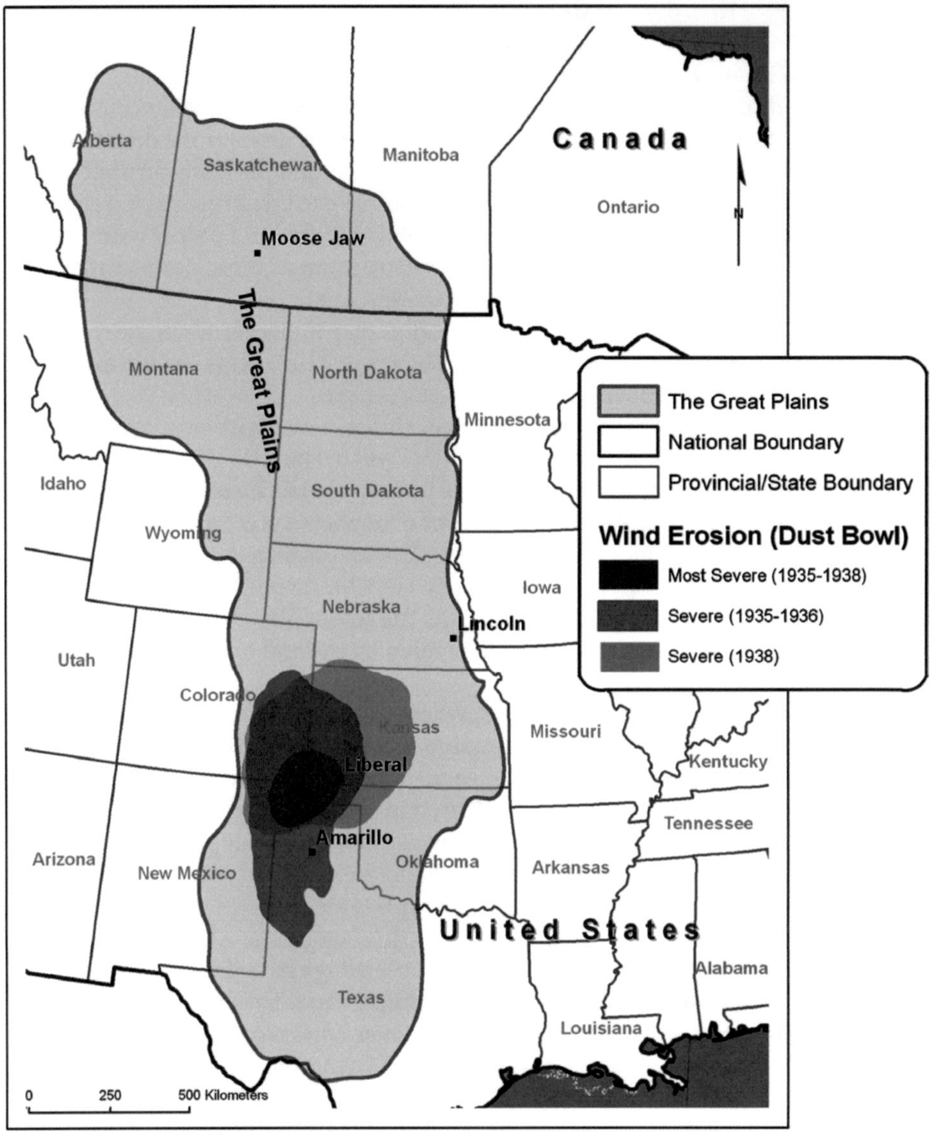

Figure 4.
The Great Plains and the dust bowl proper: Note that the most severely affected area was limited to NE New Mexico, N Texas, W. Oklahoma, SW Kansas, and SE Colorado; Figure 1 of [19].

the drought of 1855–1864 may have been the most severe and sustained *spring* moisture deficit over the Kansas-Missouri study area; and that the drought of the Dust Bowl era was by far the most severe and sustained *summer* precipitation deficit over the area. Nonetheless, when the precipitation data are summarized by growing season, the Dust Bowl drought was not remarkably more severe than the droughts of the 1860s, 1910s, and 1950s.

Perhaps the most seminal contribution of the above-summarized work is how the authors put the 1930s Dust Bowl into a much longer historical context. This context is lengthened considerably by the article summarized next.

Another research team [25] gives a much longer view of moisture/drought in the Dust Bowl area with the Palmer Drought Severity Index (PDSI). This index approximates soil moisture relative to 'normal' conditions, using meteorological data and assumptions about soil properties. 'Drought' is here defined as starting

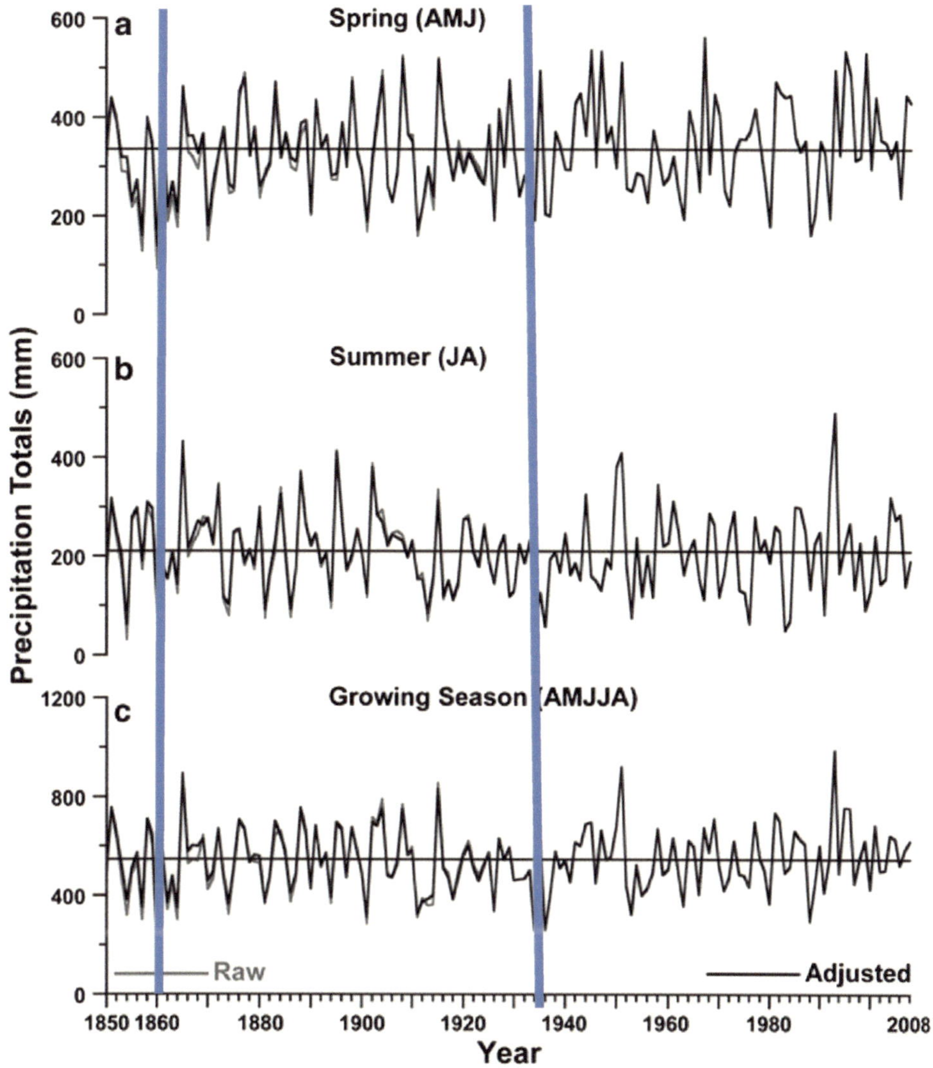

Figure 5.
Monthly precipitation from 20 Kansas and Missouri meteorological sites: Figure 2 of [24], augmented by marker lines for 1860 and 1935; AMJ, April, May, and June; JA, July, August; AMJJA, April – August. Note the difference in vertical scales: The top two go from zero to 600 mm, the bottom, from zero to 1200 mm.

the first month when the PDSI is less than −1 for three or more months and ending the month before the PDSI is greater than −1 for two or more months." **Figure 6** presents this drought index for 1,000 years in southeastern Colorado.

This prehistorical to historical reconstruction shows that this drought index sank below −3 about 12 times, with the worst (−4) and longest duration in 1470, compared with the 1935 Dust Bowl value of −2.8. This suggests two points: (1) that severe droughts occur roughly every 80 years, and (2) that the drought conditions of the Dust Bowl were severe but that other droughts have been worse. The strongest point in this research, which relies on tree ring data, is its temporal expansion from years and a century and a half to a complete millennium. In the works summarized so far, then, we go from the 1930s, to 1850–2008, and to 1000–2000 – the short, medium, and long-term views.

Yet another research approach to better understand the Dust Bowl examines changes in the land surface [26]. The authors state that "the drastic land-cover

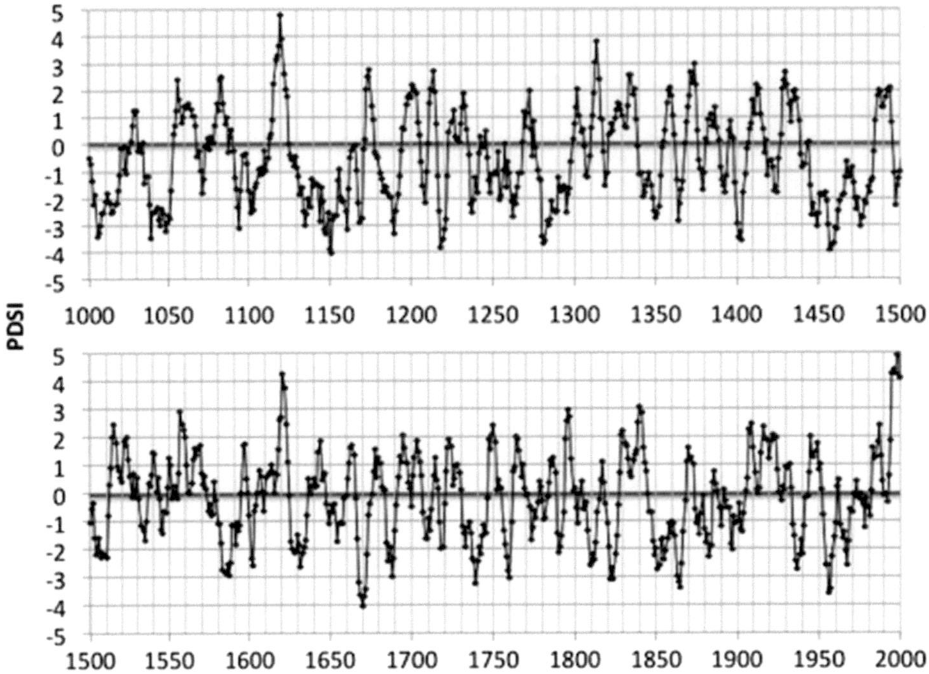

Figure 6.
Five-year moving average of palmer drought severity index values for ... southeastern Colorado ... for 1000–2000 AD. The values [come] from tree ring records. (Figure 17 of [25]).

change from pre-settlement to the 1930s in the Great Plains resulted in a strong increase in the surface albedo. ("Surface albedo" quantifies the fraction of the sunlight reflected by the Earth's surface.) On average, the albedo changes from ~0.16 in the native grassland to ~0.20 in dryland cropland, and such a change can considerably alter the surface energy budget. In [their] simulations, changes in surface albedo from pre-settlement to the 1930s land-cover resulted in a 5 Wm^{-2} reduction in solar energy absorbed at the surface (averaged over the Great Plains from May to July)." (Incoming solar radiation is often expressed as energy (Watts) per square meter (m^{-2}.) They extend this argument by explaining how these energy budget changes contributed to the 1930s drought. **Figure 7** depicts how surface albedo has changed from the 1930s to the present day.

Although the work just summarized may seem somewhat unrelated to dust emissions, it does analyze dust potential through changes in the surface land cover. Arguably, surface land cover dictates the potential for dust suspension under any given set of intense meteorological conditions. This paper would be accessible to most general readers.

In contrast to the above regional analysis, another researcher investigated the relation between meteorology and dust emissions [27] on a micro-scale for a north Texas dust storm that occurred in 1937. **Figure 8** displays the micro-geographic extent of dust emissions for a 4 km^2 sand dune area in Texas.

The authors conclude that:

1. Lower-level atmospheric and surface air temperatures are the strongest drivers of Dust Bowl dust events, followed by low relative humidity. Anomalies in this thermal gradient and moisture carried by the Great Plains Low Level Jet occurred on dust event days that were not present on days without dust events within the same season.

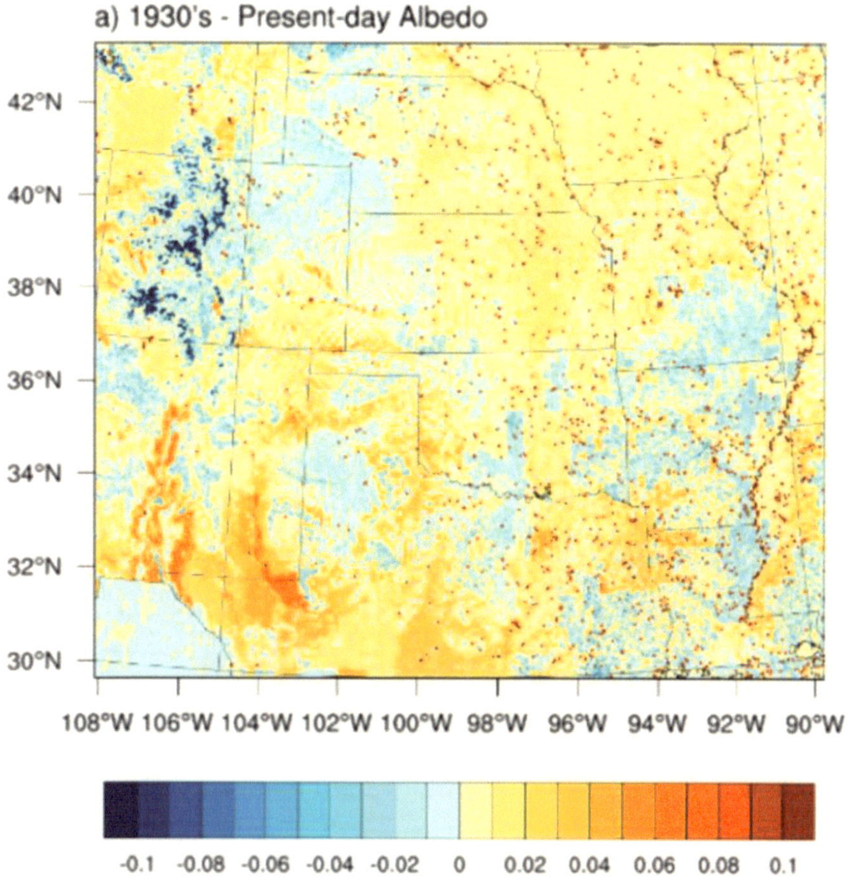

Figure 7.
1930s surface albedo minus present-day surface albedo. The tiny dots are water bodies that have distinctively different albedo from the land; Figure 3a of [26]. The differences between the two albedos are lowest in the Rocky Mountains and highest in W. Texas near El Paso.

2. Four modes of dust events were related to the season of occurrence and dominant meteorological controls. Two modes characterize "blowing season" events, with spring (MAM) dust events related to an inversion of surface and atmospheric air temperatures, and summer (JJA) dust events associated with intensified surface heating. The third mode of dust event occurs during the winter (DJF) after an extended dry period, and the fourth dust event mode reflects the passage of vigorous synoptic cold fronts that can occur in any season. (Note: [MAM = March, April, and May; JJA = June, July, and August; DJF = December, January, and February.)

3. PM_{10} emissions from common dust sources across the Southern High Plains indicates that anthropogenic disturbance of surface crusts can increase their magnitude from 0.001 to 0.01 mg m^{-2} s^{-1} from siltier soils. (The units here are milligrams (mg) per meter (m) squared, per second (s).) Emissions from loose, uncultivated sandy soils, however, can emit similarly potent levels of dust, suggesting a more complex narrative for landscape degradation in the 1930s Dust Bowl.

This paper is of moderate technical difficulty but would be understandable to most general readers. Admittedly, the authors partake of a somewhat oblique

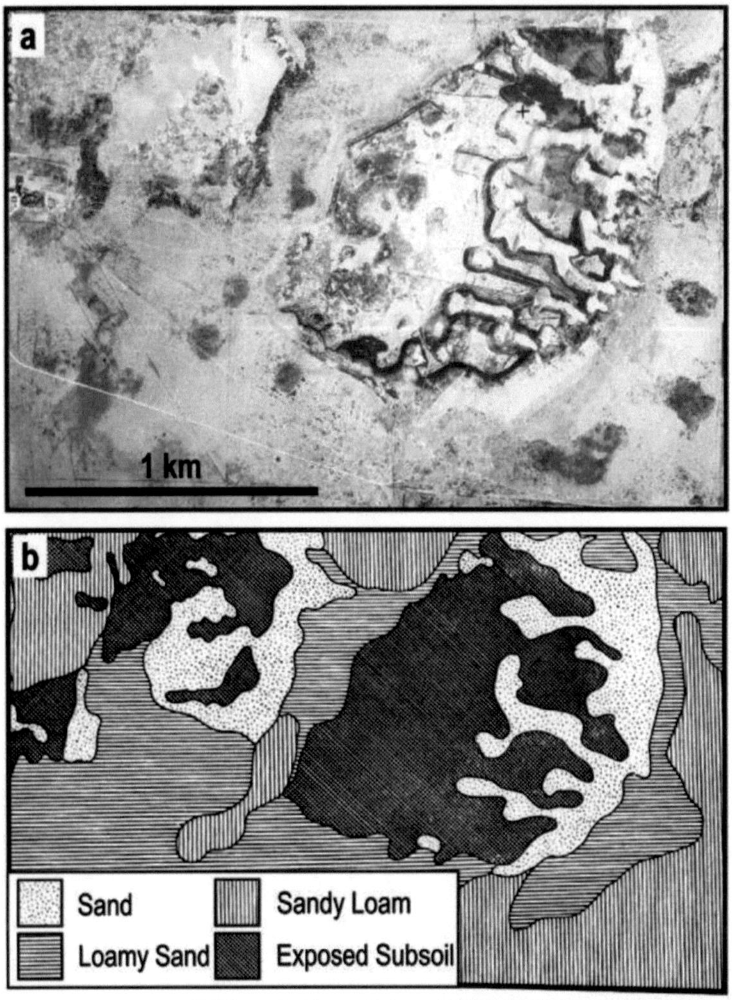

C Soil	Bare Area (m^2)	% Silt	Flux (mg m^{-2} s^{-1})	PDE (kg)
Sand	475	6.5	0.281	4.417
Loamy Sand	1,005	8	2.845	106.9
Sandy Loam	783	18.8	*Crusted:* 0.007 *Mantled:* 1.564 *Disturbed:* 5.153	*Crusted:* 0.363 *Mantled:* 76.75 *Disturbed:* 252.8
Subsoil	805	12.6	*Crusted:* 0.007 *Mantled:* 1.564 *Disturbed:* 5.153	*Crusted:* 0.294 *Mantled:* 62.26 *Disturbed:* 205.1

Figure 8.
Potential dust emissions from the Dalhart sand dune area in Dallam County, TX for an event on April 7, 1937. (a) Aerial photograph of the case study area captured on October 5, 1936. (b) Contemporaneous soil texture map produced by the soil conservation service. (c) Available bare surface area in photograph to emit dust by soil texture, the percent silt in the mapped soil unit, and the derived PM10 flux rate. (Figure 15 from [27]).

approach to dust emissions in their concentration on landscape characteristics; but, after all, landscape cover must be considered an essential, if not paramount, ingredient in the severity and frequency of dust storms.

From these more physical-science oriented summaries, this review paper now delves into the human misery of the Dust Bowl -- a tragedy of sad and immense proportions. This section relies on remarks by an Oklahoman physician, as chronicled in [10], p. 173. "In a report delivered to the Southern Medical Association [in April 1935], Dr. John H. Blue of Guymon, Oklahoma, said he treated fifty-six patients for dust pneumonia and all of them showed signs of silicosis; others were suffering early signs of tuberculosis. The doctor had looked inside an otherwise healthy farm hand in his early twenties and told him, "You are filled with dirt". The young man died the next day. The doctor then discusses silicosis, stating that prairie dust has a high silica content, and comparing the respiratory distress of Dust Bowl citizens to that of underground miners. He points out one important difference: silicosis in miners takes many years to build up, whereas doctors in the Dust Bowl were seeing a condition like silicosis after just three years of storms. For those residents who stayed the course, the human toll must have been devastating. For many of those who migrated, as depicted in John Steinbeck's *The Grapes of Wrath*, the outcome was also less than rosy. This reference [10], a clear historical, journalistic account of the Dust Bowl, would be accessible to all readers.

After the sections of this paper on present-day dune fields and the 1930s Dust Bowl, the concluding section first explains the basics of dust storms as described in two lengthy reports. Included in this discussion is a summary of how and why these storms are formed and how the public is alerted to them. Second, from two technical articles, it explains how scientific communities grapple with the difficulties of numerically simulating dust storms, work which might ultimately lead to a better predictive capacity.

4. Dust storms, their causes, effects, and attempts to forecast and model them

This third and concluding section describes both the basics of dust storms and how forecasting and numerical simulation of dust storms are accomplished today. Dust and sandstorms afflict most arid and semi-arid regions of the world, cause serious problems in commercial air traffic and vehicular traffic, degrade building surfaces, lead to increased house and office cleaning costs, and adversely affect human respiratory health. Because of their ultra-high turbulence as they contact the land surface, and because their resultant particulates concentrations are extremely heterogeneous in both time and space, accurately simulating these phenomena remains an elusive goal. Despite these shortcomings in the simulations, weather forecasters are still faced with the necessity of predicting these storms' locations, durations, and severities. These predictions then allow authorities such as the National Weather Service or highway departments to broadcast near real-time warnings to the vehicular-driving public.

As the authors live in Arizona, the next remarks concern the landscapes and weather of this state. Arizona has three distinct physiographical provinces: (1) lowland deserts in the south and southwest, (2) rugged mountainous highlands in its north-central region, and (3) the Colorado Plateau -- a broad, high- elevation plain comprising its northern third. In the populated areas the elevations range from 43 m (140 feet) above sea level in the far southwest corner at Yuma to about 2100 m (6,900 feet), an elevation that extends from the north-centrally located Flagstaff in a broad swath to the east-southeast, culminating in the far east-central region next to New Mexico. These substantial elevation differences lead to pronounced differences in weather. Extreme inclement weather often leads to vehicular crashes, whether it is heavy rain, thick fog, heavy snowfall, icy roads from wet winter rain or snow, or blowing dust. Except for fog, a rare phenomenon in Arizona, this marked spatial variation in weather leads to vehicular crashes from all these extreme weather conditions somewhere in the state. The following discussion is limited, however, to the lowland deserts and blowing dust. The focus here is on how the predictive capacity of the weather-forecasting and of the atmospheric science communities has been brought to bear on developing better dust predictions to reduce vehicular crashes and to reduce population exposure to unhealthful levels of airborne particulates. In recent years, a considerable body of work on this subject has been conducted by the National Weather Service, by the Arizona Department of Transportation, by other governmental agencies, and by academic researchers.

4.1 Dust storm and sandstorm basics

Two lengthy reports provide information for this discussion, first, a global report assembled by multiple researchers for the United Nations Environmental Programme [1], and second, a comparable report limited to Arizona [28]. Both reports thoroughly discuss the atmospheric physics and meteorology of dust storms and sandstorms. Both estimate the economic damage wrought by dust storms. Both consider mitigation efforts to reduce the flux of anthropogenic dust. To the interested reader, both are worth obtaining and studying.

Dust storms occur whenever strong winds encounter dry, erodible land surfaces. Entrainment of particles occurs when the wind shear stress exceeds the ability of the surface material to resist detachment or transport. Wind erosivity is a product of wind velocity and wind flow characteristics, especially turbulence near the ground. In addition to ambient wind speed, vegetation and land-form characteristics of surface roughness play a large role in determining wind erosivity. Local wind conditions are also influenced by wind systems generated over larger areas, and thus may depend on land use and other physical characteristics in neighboring regions. Consider the "dry, erodible land surfaces", which, according to [1], consist of 75% natural landscapes, such as the Sahara or Gobi Deserts and dry lake beds, and of 25% anthropogenic land surfaces, such as active or abandoned agricultural fields, unpaved roads, large mining and construction sites, and so forth. Mitigating dust emissions can only be directed to the anthropogenic dry land surfaces, so mitigation discussions are limited to this human-caused one fourth of the problem. The global report [1] presents their Table 2.2, p. 10, that gives the different types of land surfaces that can or cannot produce dust in high winds (**Table 3**).

Although many of these land surfaces have low or moderate dust or sand potential, the more important ones are (1) lakes that are ephemeral or are of dry, non-consolidated surfaces; (2) high-relief alluvial deposits that are both unarmored and unincised; and eolian sand dunes. Most agricultural soils for growing crops have been accumulated through alluvial processes, oftentimes augmented by the deposition of loess, so these fields when in between crops or when fallow or abandoned have high potential for dust emissions. Lakes in this list of dust potential actors are sometimes shallow water bodies constructed in large irrigation projects, but for various reasons the upstream waters that could be diverted to fill them become unavailable, leaving expansive dry lake beds prone to heavy dust emissions. In contrast with the eolian sand dunes and natural dry lake beds, these two categories of dust producers are amenable to mitigation.

Geomorphic type	Typical texture	Importance for dust emissions
Lakes		
Wet	Sand, silt, clay	Low
Ephemeral	Silt, clay	High (if sandblasting, medium)
Dry consolidated	Silt, clay	Low
Dry, non-consolidated	Silt, clay	High (if sandblasting, medium)
High relief alluvial deposits		
Armored, incised	Mega-gravel, gravel, sand	Low
Armored, unincised	Mega-gravel, gravel, sand	Low
Unarmored, incised	Gravel, sand, silt, clay	Medium
Unarmored, unincised	Sand, silt, clay	Medium-high
Low relief alluvial deposits		
Armored, incised	Gravel, sand	Low
Armored, unincised	Gravel, sand, silt, clay	Medium
Unarmored, incised	Sand, silt, clay	Low
Unarmored, unincised	Sand, silt, clay	Medium
Stoney surfaces	Gravel, sand, silt, clay	Low
Sand deposits		
Sand sheet	Sand	Low to medium
Eolian sand dunes	Sand	Low to high
Loess	Silt, clay	Low-medium
Low emission surfaces: bedrock, rocky slopes, dunecrust, snow/ice permanent cover	Mega-gravel, gravel, sand, silt, clay	Low

Table 3.
Different land surface types that can (or cannot) produce blowing sand and dust, (Table 2.2 of [1]).

The literature on landscape characteristics of dust potential, and on the atmospheric physics and meteorology of the formation, transport, and eventual dissipation of dust storms is both voluminous and can be highly technical. For greater detail, the reader is referred to the two already cited reports or other textbooks or journal articles. Both long reports would be comprehensible for the average reader.

To summarize dust storms, their *formation* comes about from extremely high surface winds, produced either from massive thunderstorms or from synoptic weather fronts. Their direction and distance of *transport* is determined by the continuing influence of these winds, in conjunction with their continued ability to contact erodible land surfaces. Their *dissipation* occurs as the wind speeds decrease with the weakening of either the thunderstorm activity or the large-scale frontal movements. The global report presents in their Figure 2.2 and Table 2.1 a clear conceptualization of the phenomenon, along with the influential physical characteristics of weather variables, of soil surfaces, of vegetation, and of landforms, shown as **Figure 9** and **Table 4**.

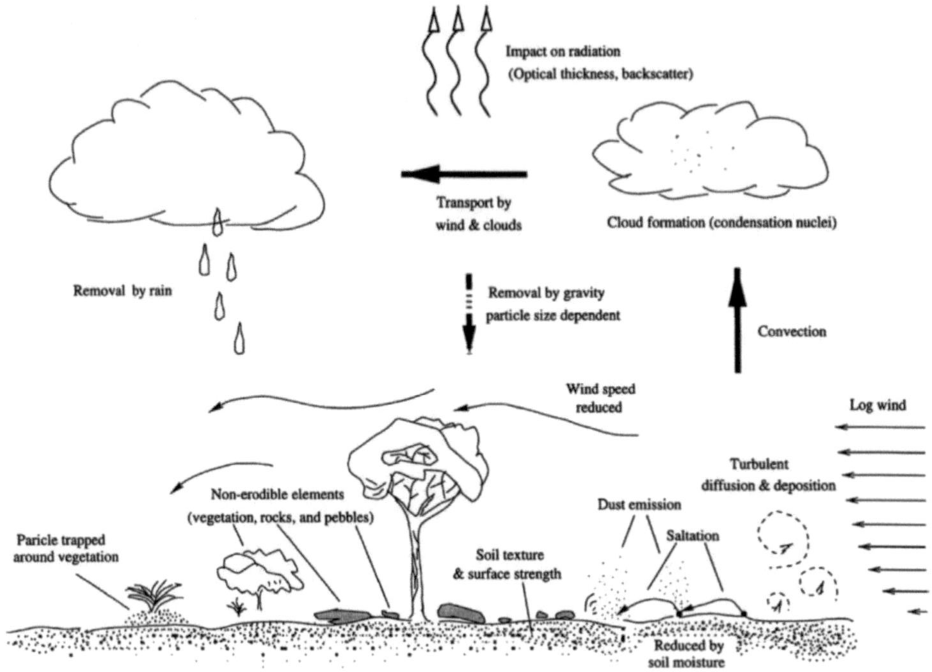

Figure 9.
Dust storm formation processes (Figure 2.2 of [1]).

Climate	Sediment or soil	Vegetation	Landform
Wind speed (+)	Soil/sediment type	Type	Surface roughness (+/−)
Wind direction	Particle composition	Coverage (−)	Slope (−)
Turbulence (+)	Soil/sediment structure	Density	Ridge
Precipitation (−)	Organic matter (+)	Distribution (+/−)	
Evaporation (+)	Carbonates (−)		
Air temperature (+/−)	Bulk density		
Air pressure (+)	Degree of aggregation (−)		
Freeze–thaw action (+/−)	Surface moisture (−)		

(+) indicates that the factor reenforces wind erosion; (−) indicates that the factor has a protective effect, reducing wind erosion; (+/−) indicates that the effect can be positive or negative depending on the processes involved.

Table 4.
Key physical factors influencing wind erosion (Table 2.1 of [1]).

In the Arizona report [28] the authors state that in Phoenix from 1948 to 2015 the number of summer dust storms ranges from three to five in the earlier years down to one to three in the latter years. They speculate that the decrease in frequency may stem from the expanding Phoenix urban area, in which formerly outlying agricultural lands with considerable dust potential have been converted into residential and commercial buildings, into landscaping that includes parks and lawns, and into what generally is called the "built environment". They present one photograph of a dust storm, shown as **Figure 10**.

Figure 10.
Dust storm of 5 July 2011, as it approaches the National Weather Service office at Sky Harbor airport in Phoenix: (Figure 36 of [28]).

As for the observational tools and methods of predicting these dust storms, the authors offer the following, summarized in **Table 5**.

That concludes the summaries of the lengthy global and Arizona reports on all aspects of dust storms. This review paper now continues, and concludes, with a discussion of how these storms can be forecast or simulated.

4.2 Forecasting and simulating dust storms

The narrative immediately above covers the day-to-day observational tools and predictive systems for dust storms. At least two questions remain: (1) how are these dust storms studied by numerical simulations, and (2) how well do these simulations match the various observations such as satellite observations, Doppler radar images, and ground-based measurements of particulates concentrations? What follows are two examples of recent research on dust storm simulations. Both examples are highly technical papers unsuited for the non-technical reader.

One instructive example of the difficulties in performing these simulations and of what improvements might be forthcoming, can be found in the work of [12]. The authors assert that "regional-to-global models generally do not accurately simulate these storms", for two reasons: "(1) using a single mean value for wind speed per grid box, i.e., not accounting for subgrid wind variability and (2) using convective parametrizations that poorly simulate cold pool outflows". Their remedies take two forms. First, they "incorporate a probability distribution function for surface wind in each grid box to account for subgrid wind variability due to dry and moist convection." Second, they use "lightning assimilation to increase the accuracy of the convective parameterization to better simulate cold pool outflows".

These researchers built the subgrid wind variability and lightning assimilation into two different physico-chemical models: the Weather Research and Forecasting

Tool, warning, or prediction system	Remarks
Low-cost air quality sensors	\$100; measures [PM_{10}] every 30 seconds; data sent to a central server
Traditional continuous particulates monitors	Operated by air pollution agencies; data can be retrieved near real-time
Human weather spotters	Trained; report blowing dust to NWS offices
Automated Surface Observing System (ASOS)	Hourly reports, data in 1- to 5-minute intervals; at all major and many smaller arizona airports
Doppler weather radar with horizontal and vertical pulses	Detects only the major storms; give 2-dimensional pictures; only three operate in Arizona
Satellite imagery	Clear depiction of large dust storms; but only two passes per day over any one area, so unlikely to capture many storms
Traffic cameras and web cams	Useful, but unable to distinguish dust storms from other smaller dust sources
Dust storm warning and wireless emergency alerts	Through existing NWS platforms and the media through the emergency broadcast system; now based on smaller areas (polygons) to avoid alerting citizens who may be 100 km away from the storm.
Electronic message signs on highways	Arizona Department of Transportation has many of these, urban and rural.
Social media, especially Twitter	Valuable for both obtaining information about dust storms and disseminating critical safety information
Safety and education	Dust safety while driving: "Pull Aside Stay Alive"
Prediction system	Global Forecast System (GFS) model and the North American Mesoscale (NAM) model, both low resolution; Arizona Regional Weather Research and Forecasting (AZ-WRF) model, a high resolution model

Note: all rows in this table are from [28], except row #2 and the last row, both added by the authors.

Table 5.
Observational tools, warnings, and prediction systems for dust storms.

Model (WRF) and the Community Multiscale Air Quality model (CMAQ). The windblown dust emissions parameterizations employed incorporate saltation bombardment (sandblasting) and a novel dynamic relation for the surface roughness length. To better estimate vegetative cover, these researchers used the fraction of absorbed photosynthetically available radiation (fPAR) from the Moderate Resolution Imaging Spectroradiometer (MODIS), which is a satellite-based instrument. Earlier work showed that the modeled airborne soil concentrations agreed quite well with observations in the spring, but that it underestimated these concentrations in summer, when convective dust storms are most frequent and most severe.

As for improving convection through lightning assimilation, the authors used the Kain-Fritsch convective scheme in WRF by activating its deep convection where lightning is observed and suppressing it where lightning is absent.

The authors went on to test their modified model on the major dust storm of 5 July 2011, which began with late afternoon severe thunderstorms near Tucson, Arizona. Cold pool outflows associated with this region of large storms moved northwest toward Phoenix, bringing with them a wall of dust extending 160 km wide and 1.5–1.8 km high. Both modifications – to the winds within the subgrids and to the deep convection scheme employed when lightning was present – enabled the simulated particulates concentrations from CMAQ to better match the measured [PM_{10}], as shown in **Figure 11**. [Note: "[PM_{10}]" is read as "concentrations of PM_{10}".]

The work just summarized is highly technical, even for atmospheric scientists who study these phenomena. Missing from this work is any explicitly numerical comparison of model-generated versus observed [PM_{10}]. While the concentration maps of **Figure 11** are illustrative, they are far from definitive. The next (and last work) summarized, which does have these explicit comparisons, is also on the technical side, but perhaps is not quite as obtuse, an opinion better left to the interested reader.

This is the work of [13], in which researchers analyzed nine dust storms in south-central Arizona with the Weather Research and Forecasting model with chemistry (WRF-Chem) at 2 km resolution. The all-important windblown dust emission algorithm was the Air Force Weather Agency model [29]. In all simulations of air pollutant concentrations, it is essential to get the emissions quantified accurately both temporally and spatially. For windblown dust emissions this goal frequently proves to be elusive because the available coverages of soil moisture, surface roughness, and vegetative cover suffer from both insufficient resolution and from temporal delays between the observations of these variables and the event itself. In this highly dynamic environment, with rainfall stochastically distributed in localized pockets, and with soil surface texture varying widely even within small areas, the uncertainties of the emitted dust flux reach unreasonable proportions. Nonetheless, one proceeds with what information one has.

In comparison with ground-based [PM_{10}] observations, this modeling system unevenly reproduces the dust-storm events. The model adequately estimates the location and timing of the events, but it is unable to precisely replicate the magnitude and timing of the elevated hourly [PM_{10}]. Furthermore, the model underestimated [PM_{10}] in highly agricultural Pinal County for two reasons. First, because it underestimated surface wind speeds and, second, because the model's erodible fractions of the land surface data were too coarse to effectively resolve the active and abandoned agricultural lands.

In Phoenix the model's performance depended on the event, with both under- and over-estimations partly due to incorrect representation of urban features. Increasing the fraction of erodible surfaces in the Pinal County agricultural areas improved the simulation of [PM_{10}] in that region. Both 24-hr and 1-hr measured

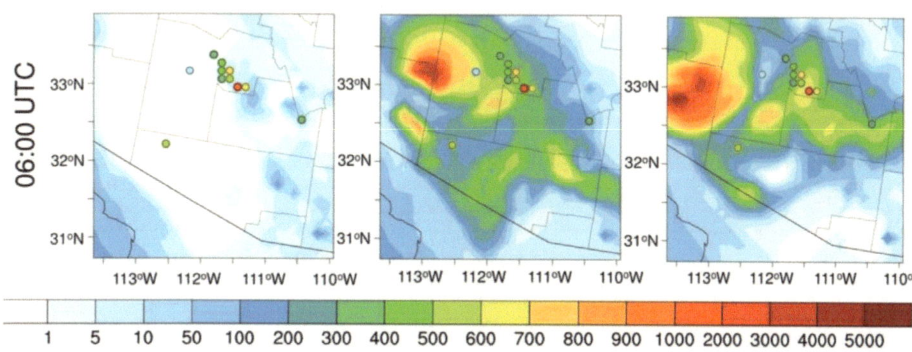

Figure 11.
Simulated hourly PM10 surface concentrations (μg m^{-3}) at 06:00 UTC on 6 July 2011 (23,00 local time on 5 July 2011) from three runs (left to right) (1) (control --no lightning assimilation (LTGA) and no subgrid wind variability (SGWV), (2) with SGWV, and (3) with SGWV and LTGA), overlaid with the observations of 11 PM10 monitoring sites. (this is lowest panel from Figure 6 of [12]).

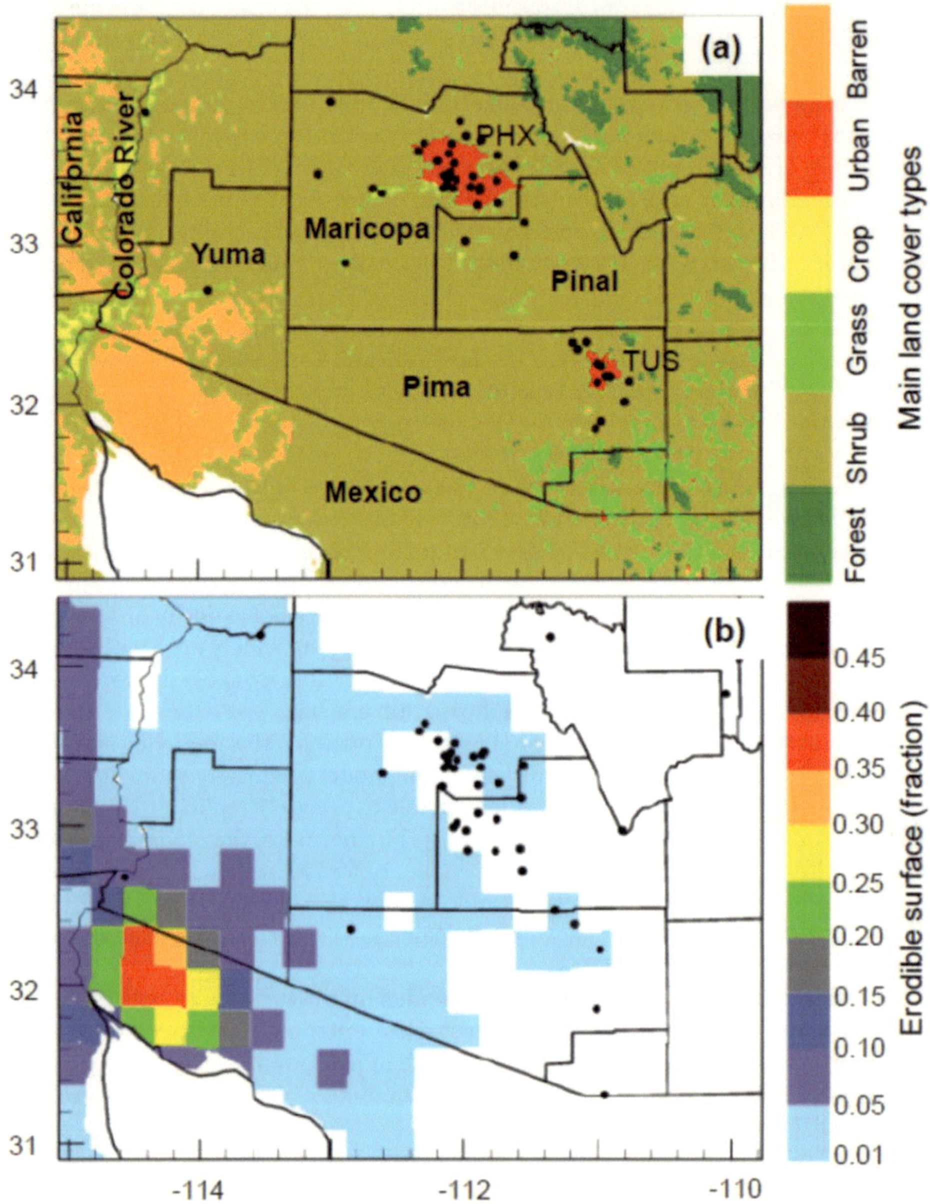

Figure 12.
Model static fields: (a) main land cover and land use type in south-Central Arizona, and (b) fraction of erodible surface (Figure 2 of [13]).

[PM_{10}] were, for the most part, and especially in Pinal County, extremely elevated, with the former exceeding the health standard by as much as 10-fold and the latter exceeding health-based guidelines by as much as 70-fold.

The authors present several graphics that depict, among other things, the landscape and the degree of erodible surface (**Figure 12**).

Figure 13 is a sample of the model's inability to match the observations, in which each panel represents a different dust storm. The observations in each case came from eight to 13 continuous PM_{10} monitoring sites, all in Pinal County. In two storms the model grossly over-estimated the observed values; in the other four the model greatly underestimated the measured peak concentrations.

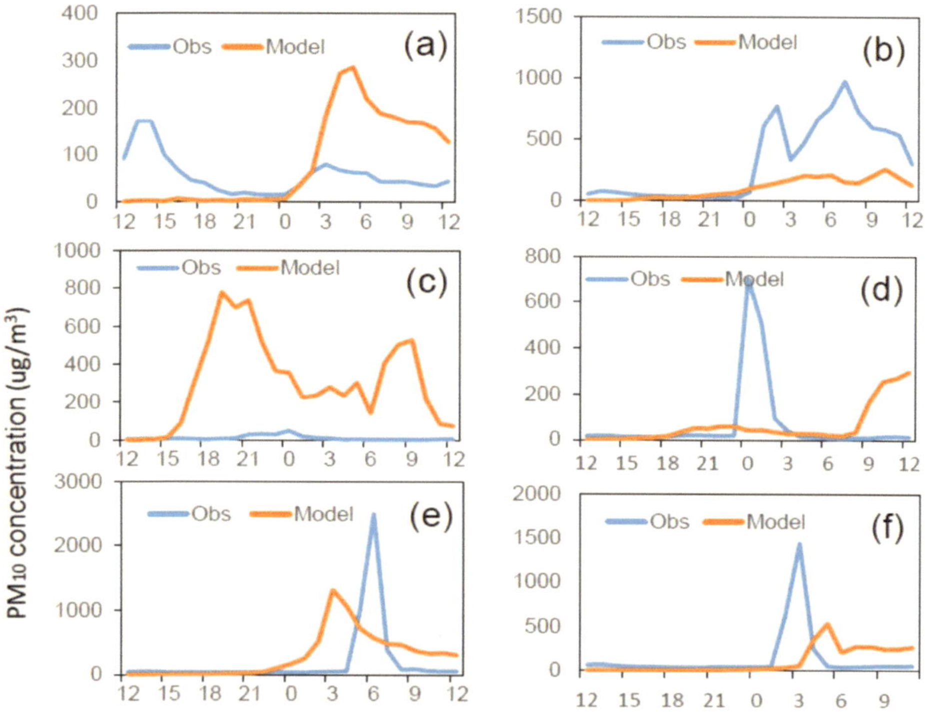

Figure 13.
Comparison of averaged PM_{10} time series over Pinal County for different cases: (a) April 13–14, 2006 (total 8 sites), (b) July 7–18, 2009 (total 9 sites), (c) January 21–22, 2010 (total 12 sites), (d) July 21–22, 2012 (total 13 sites), (e) June 30–July 1, 2013 (total 18 sites), (f) July 3–4, 2014 (18 sites), (g) June 27–28, 2015 (total 18 sites), and (h) July 7–8, 2014 (total 18 sites) (Figure 3 of [13]).

The authors conclude: "Given the severity and frequency of these dust storms and conceding that the modeling system applied did not produce the desired agreement between simulations and observations, additional research in both the windblown dust emissions model and the physico-chemical model is called for."

Thus, concludes the last part of this four-part review paper that has presented information on sandstone formation, on sand dune field formation and dynamics, on the 1930s Dust Bowl saga, and on dust storm and sandstorm basics and the forecasting and prediction thereof. The interested reader is encouraged to consult the references for a more in-depth look into these subjects.

5. Conclusions

Although sand dunes, the Dust Bowl, and forecasting and simulating dust storms may appear as three widely separated topics, they share the common bond of arising from wind-blown sand and dust. Active for millions of years and still quite active at present, this disturbing phenomenon of arid and semi-arid regions wreaks havoc with the soil, disrupts vehicular and airborne transportation, causes multiple vehicular injuries and fatalities, and degrades human respiratory health. While the geological and esthetic prospects of sand dune fields enrich the naturalists' hearts, the opposite is the case for the misery of the Dust Bowl and for the profound difficulties in predicting and simulating these dust storms. Because one quarter of these storms can be attributed to anthropogenic mismanagement of soil surfaces, it appears imperative for the agricultural and soil conservation communities to redouble their efforts at reducing the dust flux from the disturbed portions of the soil surface. Only through such concerted actions will the productivity of the agricultural fields be maintained and will the atmospheric environment be restored to a more benign equilibrium.

Acknowledgements

The corresponding author thanks the School for Engineering of Matter, Transport and Energy for their support and Professor Alex Mahalov for his.

Conflict of interest

The authors state that they have no conflicts of interest. Furthermore, the corresponding works as a non-salaried citizen's volunteer at Arizona State University and has received no funding for this work.

Abbreviations and definitions

μm	micron, one millionth of a meter, e.g. a human hair is 70 μm in diameter; the thickness of 20-pound typing paper is 100 μm
mya	millions of years ago
km	kilometer: one km equals 0.63 miles

Author details

Peter Hyde[1*] and Alex Mahalov[2,3]

1 School for Engineering of Matter, Transport and Energy, Arizona State University, Tempe, AZ, USA

2 School of Mathematical and Statistical Sciences, Arizona State University, Tempe, AZ, USA

3 Julie Ann Wrigley Global Institute of Sustainability, Arizona State University, Tempe, AZ, USA

*Address all correspondence to: peter.hyde@asu.edu

References

[1] UNEP, WMO, UNCCD, 2016, "Global assessment of sand and dust storms", United Nations environment Programme, Nairobi., Ed. G. Shepherd, UNEP. A 293-page report that documents and analyzes sand and dust storms worldwide.

[2] J. Elder, 2006, "Aeolian Dunes and Sandstone: Overview and Terminology", https://smallpond.ca/jim/sand/overview/

[3] N. Lancaster, 2009, "Monitoring aeolian features and processes", in Young, R., and Norby, L., *Geological Monitoring: Boulder, Colorado*, Geological Society of America, p. 1-25, doi: 10.1130/2009

[4] E. Dimofte et al, 20018, "A review of recent research on sand dunes formation", Researchgate, https://www.researchgate.net/publication/322356954

[5] E. Tarbuck, F Lutgens, & D. Tasa, 2016, *An Introduction to Physical Geology*,

[6] W. Norton, 2001, *The Elements of Geology*,

[7] D. Hatcher, 1994, *Structural Geology: Principles Concepts and Problems*

[8] J. Monroe, R. Wicander, and R. Hazlett, 2006, *Physical Geology: Exploring the Earth, 6th Edition*.

[9] H. Helms, 2020, *Men's Journal*, "7 Incredible Sand Dunes to Explore Around the U.S.", https://www.mensjournal.com/travel/7-incredible-sand-dunes-to-explore-around-the-us/

[10] T. Egan, 2006, *The Worst Hard Time: The Untold Story of those Who Survived the Great American Dust Bowl*, Houghton Mifflin Company, Boston

[11] National Park Service, 2021, "Natural History Handbook: Saguaro" http://npshistory.com/handbooks/natural/4b/index.htm, accessed April 2021

[12] H. Foroutan and J. Pleim, "Improving the simulation of convective dust storms in regional-to-global models", J Adv Model Earth Syst., (2017), **9**(5): 2046-2060. doi:10.1002/2017MS000953

[13] P. Hyde, J. Li, and A. Mahalov, "Simulating the meteorology and PM_{10} concentrations in Arizona dust storms using the Weather Research and Forecasting model with Chemistry (WRF-Chem)", *Journal of the Air & Waste Management Association*, (2018), **68**:3, 177-195, DOI:10.1080/10962247.2017.1357662

[14] Wikipedia, 2020, "Dust Storm", https://en.wikipedia.org/wiki/Dust_storm

[15] History Channel, 2009, "Dust Bowl", https://www.history.com/topics/great-depression/dust-bowl

[16] J. Malin, 1946, "Dust storms: Part one, 1850-1860", Kansas Historical Quarterly, **14**, 2, 129 - 133

[17] J. Malin, 1946, "Dust storms: Part two, 1861-1880", Kansas Historical Quarterly, **14**, 3,265 – 3,296

[18] J. Malin, 1946, "Dust storms: Part three, 1881-1900", Kansas Historical Quarterly, **14**, 4,391 – 4,413

[19] R. McLeman, J. Dupre, L. Ford, J. Ford, K. Gajewski, and G. Marchildon, 2014, "What we learned from the dust bowl: Lessons in science, policy, and adaptation", Popul Environ, **35**, 417-440, doi 10.1007/s11111-013-0190-z

[20] Wikipedia, 2020, "Algodones dunes", https://en.wikipedia.org/wiki/Algodones_Dunes

[21] K. Havholm and G. Kocurek, 1988, "A preliminary study of the dynamics of a modern draa, Algodones, southeastern California, USA", Sedimentology, **35**, 649-669

[22] Wikipedia, 2020, "The Gran Desierto de Altar", https://en.wikipedia.org/wiki/Gran_Desierto_de_Altar

[23] G. Blount and N. Lancaster, 1990, "Development of the Gran Desierto sand sea,of northwestern Mexico", *Geology*, **18**, 724-728

[24] D. Burnette and D. Stahle, 2013, "Historical perspective on the dust bowl drought in the Central United States", Climatic Change, **116,** 479-494, doi 10.1007/s10584-012-0525-2

[25] J. Lee and T. Gill, 2015, "Multiple causes of wind erosion in the dust bowl", Aeolian Research, **19,** 15-36

[26] Q. Hu, J. Torres-Alavez,, and M. van den Broeke, 2018, "Land-cover change and the "dust bowl" drought in the U.S. Great Plains", Journal of Climate, **31,** 4657 – 4667, doi: 10.1175/JCLI-D-17-0515.1

[27] K. Bolles, M. Sweeney, and S.Forman. 2019, "Meteorological catalysts of dust events and particle source dynamics of affected soils during the 1930s dust bowl drought, southern High Plains, USA", Anthropocene, **27,** 1 – 20, 100216

[28] G. Lader, A. Raman, J. Davis, and K. Waters, "Blowing dust and dust storms: one of Arizona's most underrated weather hazards", (2016) NOAA Technical Memorandum NWS-WR 290

[29] S. Jones and G. Creighton, 2011, "Air Force Weather Agency dust emission scheme for theWRFchem/GOCART",http://www2.mmm.ucar.edu/wrf/users/workshops/WS2011/Power%20Points%202011/8A_1_Jones_WRFWorkshop_11.pdf

Chapter 2

Ecological Effects of Oasis Shelterbelts in Ulan Buh Desert

Fengmin Luo, Zhiming Xin, Junliang Gao, Yuan Ma, Xing Li and Huaiyuan Liu

Abstract

In arid region, shelterbelt is the ecological barrier for oasis. Understanding its ecological effects can provide theoretical supports for its long-term management and sustainable development. Two standard meteorological stations were used to monitor climatic factors continuously for 7 years, and two 50 m dust monitoring towers were used to continuously monitor sandstorm for 10 times, which were located inside and outside oasis shelterbelts in the northeastern edge of Ulan Buh Desert. The microclimate differences were analyzed, as well as the ecological effects of oasis shelterbelts was clarified inside and outside oasis. In the present study, under the influence of a large-scale shelterbelts, air temperature, land ground temperature and evaporation respectively decreased 5.13% ~ 24.74%, 2.38% ~ 20.09% and 7.06% ~ 17.68%, whereas the relative humidity and precipitation respectively increased 6.93% ~ 25.53% and 4.30% ~ 50.15%. During the occurrence of sandstorms, the wind speed inside and outside shelterbelt showed an increasing trend with the increase in height. The relationship between wind speed and height was expressed as a power function. The wind direction was mainly W, WNW and NE, but the proportion of each direction was different inside and outside shelterbelt. When the sandstorm passed through oasis shelterbelts, the wind speed was significantly weakened, with an average reduction of 30.68%. The horizontal aeolian sediment flux decreased 414.44 $g{\cdot}m^{-2}$ and the aeolian deposition flux decreased 0.81 $g{\cdot}m^{-2}$. The results revealed that the microclimate was improved by oasis shelterbelts, especially in the growing season. Therefore, oasis shelterbelts help to maintain the sustainable development of oasis.

Keywords: meteorological factor, sandstorm, oasis shelterbelts, Ulan Buh Desert

1. Introduction

Oasis is a unique geographical landscape in arid and semi-arid regions, which plays an extremely role in the development of human society [1, 2]. In arid region, oasis is irreplaceable in landscape, environment and function. Climate change is an important driving factor affecting ecosystem services. Rising temperature, precipitation change, and extreme climate events will have a great impact on the ecosystem [3]. Climate resources are the most important renewable resources for the development of oasis. Meanwhile, oasis will have a significant impact on climate change [4]. Sandstorm will cause long-term climate effects [5]. Wind is an indispensable dynamic condition for the occurrence of sandstorm, and it is also the main indicator for determining sandstorm intensity [6]. Dust monitoring is a necessary means to provide early warning of sandstorm, to study prevention and control measures, and to reduce the damage caused by sand-dust disasters [7]. The near-surface layer (0 ~ 50 m) not only provides the source of dust, but also the main space for people's life, industrial and agricultural production. Therefore, strengthening the monitoring of near-surface dust and making early warning has important practical significance in disaster prevention.

Ulan Buh Desert is one of important sand sources in China, which has serious wind and sand disasters. It is located at 106°09′–106°57′ E and 39°16′–40°57′ N, the arid region of northwest China, covering nearly 11,000 km^2 [8]. The desert lies at an elevation from 1028 to 1054 m [9], where the southwest area is topographically higher than the northeast [10]. Geomorphologically, the types of sand dunes in the area include moving, fixed, and semi-fixed dunes, the proportions of which are almost equal [11]. The desert is expanding to east and south, which affect the normal functions of the transportation network and water conservancy facilities in its territory, threaten the ecological security of the oasis, the development of agriculture and animal husbandry, and the health of residents [12, 13]. As an important part of the "Three-North Shelter Forest Program", Oasis in Ulan Buh Desert plays an important role in promoting the economic development of the Hetao region and reducing wind and sand disasters. Many studies have been carried out in windbreak structure and configuration of shelterbelts, even effectiveness of shelterbelt [14, 15], dust reduction mechanism and effect of sand reduction [16–18]. In recent years, the impact of meteorological elements (wind speed, precipitation and temperature, etc.) and underlying surface elements (vegetation coverage, soil moisture, degree of surface consolidation or looseness and sediment size) on sandstorms were studied worldwide, including Mojave Desert, Junggar Basin, North and Northwest China [19–22]. However, there were relatively few systematic studies on the ecological effects of oasis shelterbelt systems [23].

In arid region, oasis is a necessary condition for economic development, while the climatic factors play a vital role in maintaining the sustainable development of oasis [24]. Therefore, it is particularly important to explore the ecological effects of oasis shelterbelt system in the Ulan Buh Desert. Therefore, the annual dynamic of climate was studied by meteorological data from 2012 to 2018, which came from two meteorological stations inside and outside oasis shelterbelts in Ulan Buh Desert. Meanwhile, sandstorm was analyzed by the wind speed, wind direction and dust flux data from 10 sandstorms of 50 m dust monitoring tower. Therefore, the differences and causes of microclimates are discussed inside and outside the oasis shelterbelts. And effectiveness of shelterbelts is explained. The results can help to predict the trend of microclimate changes precisely in the future, reveal the characteristics of low-altitude sandstorm on the northeastern edge of Ulan Buh Desert, and give theoretical foundation for the management and sustainable development of the oasis shelterbelts.

2. Materials and methods

2.1 Study area

The study area is located in the northeast edge of Ulan Buh Desert. It belongs to Dengkou County, Inner Mongolia. The region has a temperate continental monsoonal climate, which are affected by the southeast monsoon in summer and autumn, and are controlled by the Siberian-Mongolian cold anticyclone in winter and spring. The mean annual air temperature was 7.8°C [9], with highest air temperature of 25.6°C in July and lowest air temperature of −11.5°C in January. The mean annual precipitation was 140.0 mm, primarily distributed in the summer. However, mean annual potential evaporation was 2372 mm. The mean annual wind speed was 3.7 m/s with the mean annual windy days of 10–32 d. The frost free days were 168d. The annual sunshine time was 3229.9 h [25]. The soil types are mainly eolian soil and sandy loam soil. The natural vegetation is dominated by *Nitraria tangutorum*, *Artemisia ordosica*, *Calligonum alaschanicum* and *Haloxylon ammodendron* [26]. The dominant tree of shelterbelt is *Populus alba var. pyramidalis* in oasis. The distance between two observation sites is 2.85 km. There are fixed and semi-fixed dunes outside shelterbelt, which dominated by *Nitraria tangutorum* with the height of 1.2 ~ 3.6 m. The area of the shelterbelt system is 1487.3 ha. The width of a belt is 32 m, which is composed of 8 rows of trees, with the auxiliary forest belt spacing of 98 m, and the miniature forest belt spacing of 398 m. The characteristics and locations of vegetation inside and outside shelterbelt are shown in **Figure 1**.

2.2 Data sources and research methods

2.2.1 Meteorological factors

The parallel comparative experimental observation of meteorological elements inside and outside oasis shelterbelt were conducted by Windsonic two-dimensional ultrasonic wind speed and direction sensors (1590-PK-020, Campbell, USA) and temperature and humidity sensors (1590- PK-020, Campbell, USA). The data was collected from January 1, 2012 to December 31, 2018. The start wind speed of wind speed and direction sensor is 0.01 $m{\cdot}s^{-1}$, with the accuracy of ±2% and wind direction of ±3°, which range from 0 ~ 60 $m{\cdot}s^{-1}$ and 0 ~ 359°, with the resolution of 0.01 $m{\cdot}s^{-1}$ and 1°, respectively. The precipitation and evaporation were collected in accordance with the relevant regulations of the "Ground Meteorological Observation Regulations" issued by the China Meteorological Administration [27]. The observation time was 8:00, 14:00, and 20:00 (Beijing time) every day. The data at Site 1 and Site 2 were observed simultaneously and recorded instantly inside and outside shelterbelt, respectively. Refer to the definition of high temperature defined by the China Meteorological Administration, the temperature ≥ 35°C is a high temperature weather, and the number of high temperature days is the number

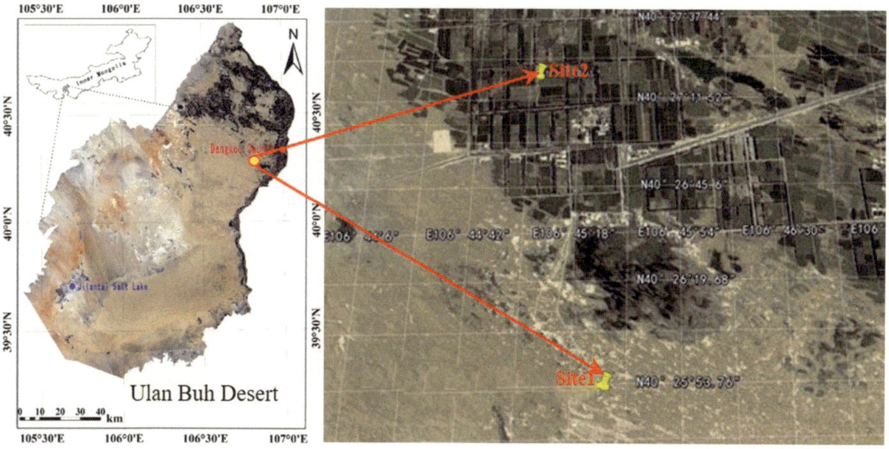

Figure 1.
The location map of the observation plot.

of days with the highest temperature ≥ 35°C. The high temperature data was the daily maximum temperature from 2012 to 2018, and the statistical period was 08:00–08:00 (Beijing time).

Temperature and relative humidity were obtained at a height of 1 m. Wind speed and direction was obtained at a height of 12 m. And precipitation and water surface evaporation were obtained at a height of 10 m. The quality of data was controlled by the simultaneous calibration of two observation points, the logical extreme value check, and the non-conformance time consistency check. According to the dividing method of four seasons in Chinese meteorology [28], spring was defined from March to May, summer was defined from June to August, autumn was defined from September to November, and winter was defined from December to February. The average value of each season's environmental factors was analyzed by Excel 2016. The variation trend graph of each meteorological factor was drawn by Origin 2017.

2.2.2 Dust collection device

The height of dust monitoring tower is 50 m, with horizontal eolian dust samplers and dust deposition traps respectively mounted at 18 heights of 0.50, 1.0, 2.0, 4.0, 6.0, 8.0, 10.0, 12.0, 16.0, 20.0, 24.0, 28.0, 32.0, 36.0, 40.0, 44.0, 48.0 and 50.0 m above the ground (**Figure 2**). The horizontal eolian dust sampler and dust deposition traps were developed by the Key Laboratory of Desert and Desertification, Chinese Academy of Sciences [29]. The horizontal eolian dust sampler was used to track the variation of wind direction and collect sand and dust from a dust inlet with the size of 20 mm × 50 mm. A dust deposition trap was designed to collect eolian sediment at different heights. This trap is a transparent glass container with an Inner diameter of 15 cm and a depth of 30 cm. There are a wind speed and direction sensor mounted at each height of 1.0, 2.0, 4.0, 8.0, 12.0, 16.0, 24.0, 36.0 and 48.0 m above the ground.

The monitor of eolian sediment flux was completed within 1 day after the end of the sandstorm. In order to ensure that the collected samples are natural air-drying, precipitation should be avoided during the sampling process. The collected samples were weighed by an electronic balance with the accuracy of 0.001 g. The 10 times sandstorms were monitored from January 2017 to June 2020. The wind speed and direction were automatically recorded during this period, and the data collection frequency was 10 min.

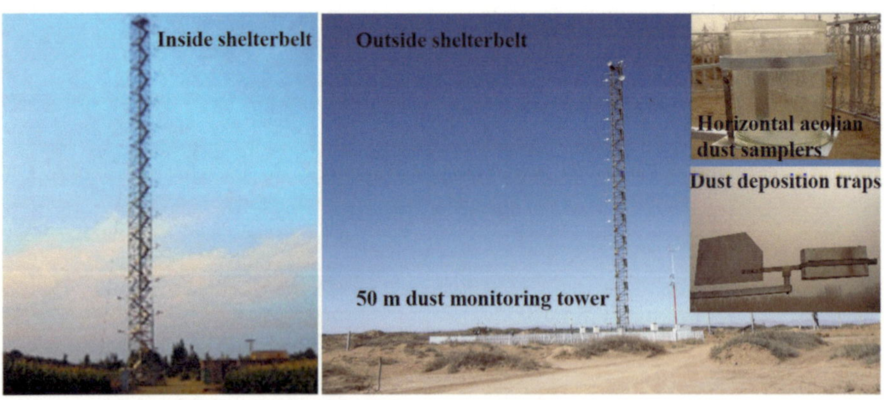

Figure 2.
The 50 m dust monitoring towers inside and outside shelterbelt, the 18 horizontal eolian dust samplers and 18 dust deposition traps.

2.2.3 Data calculation method

The horizontal aeolian sediment was calculated by formulas (1):

$$M_H = W_H / ab \tag{1}$$

Where M_H is the horizontal aeolian sediment flux (g·m^{-2}); W_H is the net weight of dust collected in the horizontal aeolian sediment samplers (g); *a* is the width of the horizontal aeolian sediment samplers opening (mm); *b* is the height of the horizontal aeolian sediment samplers opening (mm).

The aeolian deposition flux was calculated by formulas (2):

$$M_V = W_V / \pi r^2 \tag{2}$$

Where M_V is the amount of aeolian deposition fluxes (g·m^{-2}); W_V is the net weight of dust received in the aeolian deposition traps (g); *r* is the radius of the aeolian deposition traps opening (cm).

3. Results and analysis

3.1 Interannual variation of microclimate inside and outside shelterbelt

Generally, air temperature, ground temperature, evaporation, and wind speed inside shelterbelt were lower than those outside shelterbelt, while the relative humidity and precipitation were higher than those outside shelterbelt (**Table 1**). The interannual temperature outside shelterbelt varies more than inside shelterbelt, the interannual temperature outside shelterbelt fluctuates between 1.14 ~ 3.21°C, and the interannual temperature inside shelterbelt fluctuates between 0.40 ~ 0.67°C. The evaporation showed a downward trend inside shelterbelt, and it showed an increasing trend outside shelterbelt, but the changes were relatively gentle. The land ground temperature showed an overall upward trend inside and outside the shelterbelt. The precipitation showed a decreasing trend inside and outside shelterbelt, and the changes were relatively gentle. The evaporation showed a downward trend inside and outside shelterbelt. The changes of wind speed were relatively gentle inside and outside shelterbelt. Under the influence of a large-scale oasis shelterbelts, air temperature, land ground temperature and evaporation decreased 5.13% ~ 24.74%, 2.38% ~ 20.09% and 7.06% ~ 17.68%, respectively, whereas relative humidity and precipitation increased 6.93% ~ 25.53% and 4.30% ~ 50.15%, respectively.

3.2 Interannual variation of high temperature days inside and outside shelterbelt

The number of high temperature days (maximum temperature ≥ 35°C) inside and outside shelterbelt from 2012 to 2018 showed a fluctuating upward trend, and the number of high temperature days outside shelterbelt was significantly higher than that inside shelterbelt (**Figure 3**). The annual average high temperature days was 22 days outside shelterbelt, and it was 6 days inside shelterbelt. The number of high temperature days was the most, with 34 days and 17 days inside and outside shelterbelt in 2017, and the fewest high temperature days in 2012 and 2015, with 11 days and 1 days inside and outside shelterbelt, respectively. It can be seen from **Figure 4** that the high temperature days were 43 days and 154 days, the average

	Air temperature (°C)		Evaporation (mm)		Ground temperature (°C)		Precipitation (mm)		Atmospheric humidity (%)		Wind speed ($m \cdot s^{-1}$)	
Years	**Outside shelterbelt**	**Inside shelterbelt**	**Outside shelterbelt**	**Inside shelterbelt**	**Outside shelterbelt**	**Inside shelterbelt**	**Outside shelterbelt**	**Inside shelterbelt**	**Outside shelterbelt**	**Inside shelterbelt**	**Outside shelterbelt**	**Inside shelterbelt**
2012	10.8	9.1	2343.9	1987.2	25.7	21.4	193.7	229.5	49.5	62.1	3.1	1.2
2013	11.4	9.2	2333.4	2168.7	26.5	23.0	55.0	59.1	43.6	43.6	3.7	2.9
2014	12.3	9.2	2465.0	2081.7	27.5	23.9	75.	95.2	49.1	52.5	3.4	2.6
2015	10.8	9.0	2287.1	2000.6	26.6	24.2	97.9	147.0	53.6	56.5	3.3	2.7
2016	10.6	9.1	2352.8	2052.8	26.0	23.9	170.1	189.2	44.1	48.9	3.5	2.2
2017	10.2	9.3	2317.0	1925.9	27.7	26.3	86.0	89.7	34.4	38.6	3.5	2.3
2018	9.1	8.6	2403.9	1978.8	27.2	26.6	112.9	119.8	36.3	41.3	3.9	2.6

Table 1.
Annual mean of meteorological factors inside and outside shelterbelt.

maximum temperature were 36.30°C and 36.54°C, whereas the highest temperature fluctuated between 35 ~ 39.5°C and 35 ~ 43.3°C respectively inside and outside shelterbelt from 2012 to 2018. The average maximum temperature outside shelterbelt was 0.66% higher than that inside shelterbelt, but the highest temperature outside shelterbelt was 9.62% higher than that inside shelterbelt. The high temperature days were concentrated from June to August inside and outside shelterbelt, which was also the growing season of plants. The research results showed that oasis shelterbelt

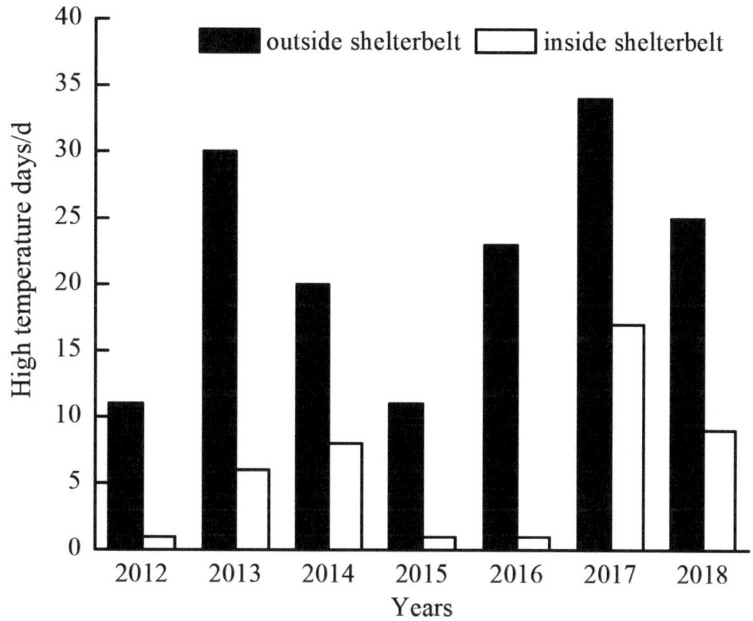

Figure 3.
Annual high temperature days inside and outside shelterbelt.

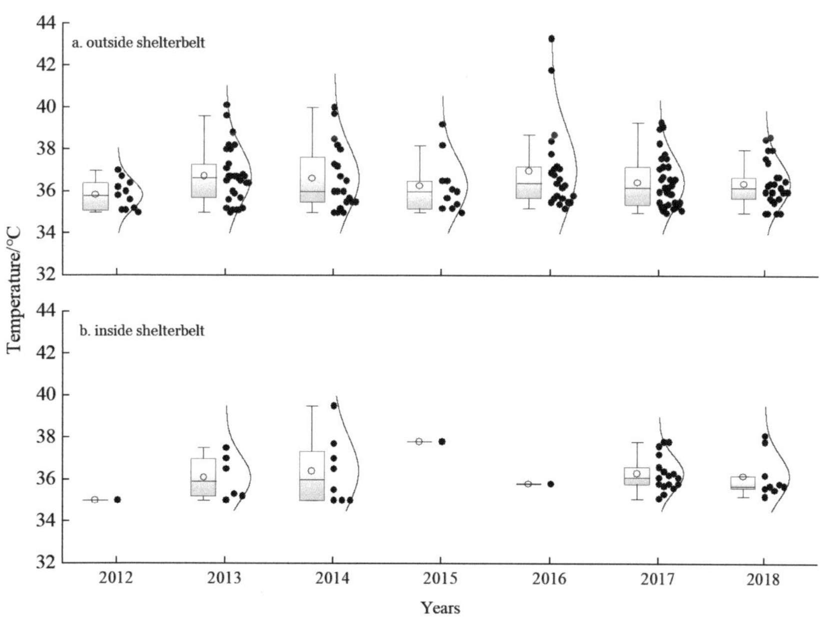

Figure 4.
The distribution characteristics of high temperature inside and outside shelterbelt.

has a good protection effect against extreme weather, it also can protect crops from high temperature damage inside shelterbelts and play a vital role in plant growth and increase crop yields.

3.3 Aeolian sediment flux

3.3.1 Wind rose diagram inside and outside shelterbelt

During the occurrence of sandstorms, the main wind directions were W, WNW, and NE, but their proportions were different inside and outside shelterbelt (**Figure 5**). Three directions inside shelterbelt accounted for 21.77%, 20.20%, and 18.00%, respectively. However, three directions outside shelterbelt accounted for 40.11%, 22.81%, and 11.69%, respectively. The wind direction inside and outside shelterbelt was similar, but the wind flow near surface ground was changed by shelterbelt. The proportion of sand driving wind in W direction outside shelterbelt was relatively high, while the proportions of other three wind directions inside shelterbelt were relatively even. Moreover, wind speed was decreased significantly by shelterbelt. The frequencies of sand driving wind exceeding 11 m·s^{-1} were 40.06% and 6.10% outside and inside shelterbelt, respectively, which was reduced by 84.77%. The frequencies of sand driving wind between 9 m·s^{-1} and 11 m·s^{-1} were 16.72% and 16.60% outside and inside shelterbelt, respectively. However, the wind speeds of sand driving wind between 5 m·s^{-1} and 7 m·s^{-1} together with those between 7 m·s^{-1} and 9 m·s^{-1} inside shelterbelt were both higher than outside shelterbelt.

3.3.2 Horizontal aeolian sediment flux

The horizontal sediment flux outside shelterbelt decreased significantly with the increase of height (**Figure 6**), and the relationship was expressed as $M_H = 1224.8\,h^{-0.343}$ ($R^2 = 0.93$, $P < 0.01$). The horizontal sediment flux inside shelterbelt increased slowly with the increase of height, which ranged from 99.62 g·m^{-2} to 280.90 g·m^{-2}. The relationship between horizontal sediment flux and height was expressed as $M_H = 130.08e^{0.023h}$ ($R^2 = 0.85$, $P < 0.01$).

The horizontal sediment flux decreased when sandstorm passed through shelterbelt. The average horizontal sediment flux within 0 ~ 50 m was 592.87 g·m^{-2} outside shelterbelt during one sandstorm, whereas the average horizontal sediment flux was only 178.43 g·m^{-2} inside shelterbelt. The concentration of horizontal

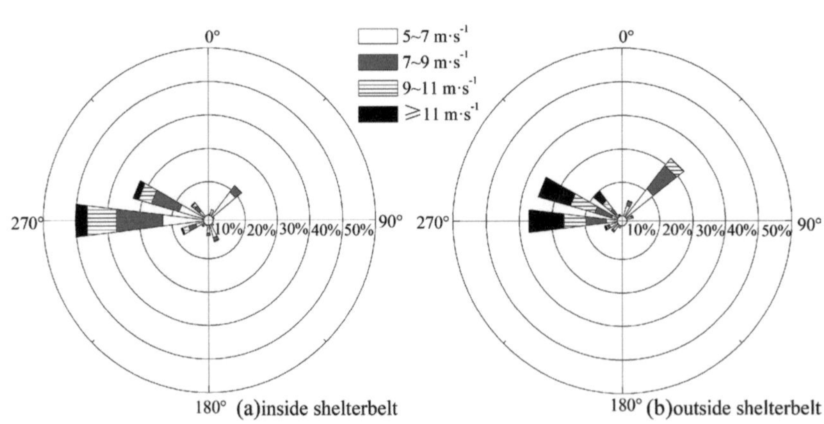

Figure 5.
Sand driving wind rose inside and outside shelterbelt during occurrence of dust storms.

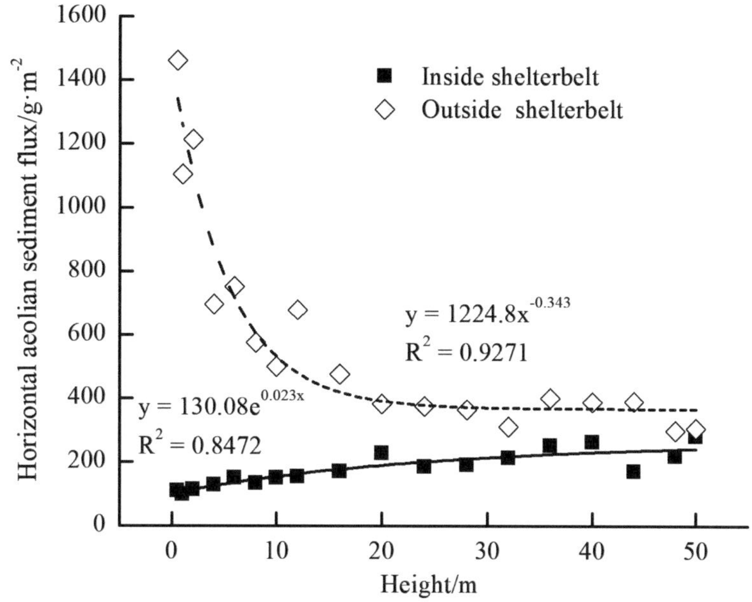

Figure 6.
Horizontal aeolian sediment flux profile inside and outside shelterbelt.

sediment flux was reduced by 69.90%. As the height increased, the difference between two sites gradually decreased with a trend of gradual overlap.

3.3.3 Aeolian deposition flux

The aeolian deposition flux decreased significantly with the increase of height inside and outside shelterbelt (**Figure 7**). Their relationship was expressed as $M_V = 6.64\,h^{-0.42}$ ($R^2 = 0.96$, $P < 0.01$) outside shelterbelt and $M_V = 4.42\,h^{-0.40}$

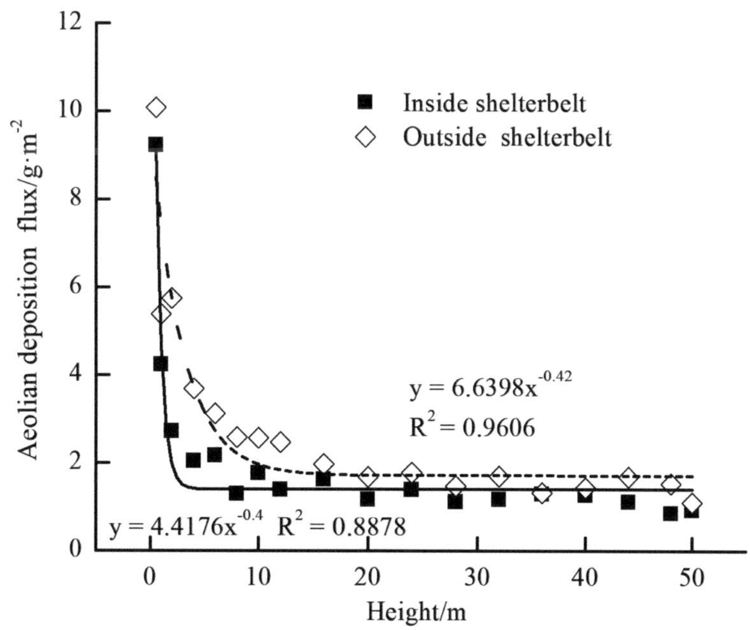

Figure 7.
Aeolian deposition flux velocity profile.

($R^2 = 0.89$, $P < 0.01$) inside shelterbelt. The aeolian deposition flux below 24 m decreased gradually with height inside and outside shelterbelt. The aeolian deposition flux above 24 m continued to decrease gradually outside shelterbelt. However, the aeolian deposition flux gradually increased above 24 m inside shelterbelt. The average aeolian deposition flux during one sandstorm within 50 m was 2.85 $g \cdot m^{-2}$ outside shelterbelt, while it was only 2.04 $g \cdot m^{-2}$ inside shelterbelt. Moreover, the difference between the aeolian deposition flux inside and outside shelterbelt also gradually decreased as the increase of height with a trend of gradual overlap.

4. Discussion

Oasis is an important area that guarantees the normal life and production of residents in arid region. Its climate effects are vital important to the sustainable development of the oasis shelterbelts [30]. The stability of the microclimate was maintained and natural disasters were reduced by oasis shelterbelts [31]. Our results showed that under the influence of a large-scale oasis shelterbelts, air temperature, ground temperature and evaporation decreased 5.13% ~ 24.74%, 2.38% ~ 20.09% and 7.06% ~ 17.68%, respectively. However, relative humidity and precipitation increased 6.93% ~ 25.53% and 4.30% ~ 50.15%, respectively. The cold-humid effect of the atmosphere inside shelterbelt formed a cold-humid column. The warm-dry effect of the desert atmosphere outside shelterbelt caused a warm-dry air current. The interaction of the cold-humid effect and the warm-dry effect formed a local circulation. The warm air outside shelterbelt was transported to the sky above the shelterbelt, thus formed a stable thermal inversion layer. Therefore, the cold and humid air inside shelterbelt was maintained, which was beneficial to crop growth in oasis [32].

The microclimate of oasis shelterbelts was conducive to the overwintering of plants and kept them from the damage of high temperature in summer. Therefore, it played a vital role in plant growth, nutrient accumulation and quality improvement [33]. The relative humidity was increased 0.5% ~ 18.6%, whereas the evaporation was decreased 18.4 ~ 1282.8 mm by oasis shelterbelts in the northeastern edge of Ulan Buh Desert. This played a positive role in increasing soil moisture and inhibiting crop transpiration, thereby increasing crop yields and improving the soil quality in long time [33]. The oasis shelterbelts in the northeastern edge of Ulan Buh Desert also had the effect of cooling in summer and heat preservation in winter. It decreased air temperature in spring, summer and autumn, whereas increased air temperature in winter inside shelterbelt. This was similar to the microclimate effect of the shelterbelt in Heihe River Basin [34]. The air heat transfer was affected by the barrier of shelterbelt, caused the temperature dynamic lag between inside and outside shelterbelt [33]. The attraction and reflection of plants inside shelterbelt to the solar radiation energy reduced the solar radiation energy absorbed by the air. Meanwhile, the growth and transpiration of plants consumed a lot of heat energy. In addition, the shading effect of the shelterbelt also resulted in the cooling effect especially in spring, summer and autumn. On the other side, it had a heat preservation effect in winter [35]. During the growing season of crop, the blocking effect on airflow was enhanced by the shelterbelt in the leafy season. The wind speed was reduced by the weakening of turbulence exchange, whereas the wind speed was increased inside shelterbelt in the leafless season [35]. Saturated water vapor was formed when the temperature inside shelterbelt was lower than that outside shelterbelt. The canopy blocked the exchange of airflow between inside and outside shelterbelt. In addition, the water vapor diffusion from inside to outside shelterbelt was reduced by the decrease of wind speed, which resulted in a higher relative humidity inside than outside shelterbelt [36].

Wind speed increased gradually with the increase of height inside and outside shelterbelt with the significant relationship of a power function. There were obvious differences in wind speed, wind direction, and eolian sediment flux inside and outside shelterbelt in the northeastern edge of Ulan Buh Desert. The results were consistent with the wind speed distribution characteristics of Minqin Oasis and Badain Jaran Desert on the oasis and desert-oasis region, which increased with the height from 0 to 50 m [37]. During the occurrence of sandstorms, wind speed profile and the migration process of sand and dust particles were affected by the type of underlying surface, thereby affecting the distribution of aeolian sediment flux [38–40]. For example, the relationships between the near-surface aeolian sediment flux and height were expressed as an exponential function in Tengger Desert [40–42], a power function in Minqin Oasis [38], whereas both functions in Taklimakan Desert [43]. Wind speed and the aeolian sediment flux through the shelterbelt were reduced significantly by oasis shelterbelts. The microclimate was improved by oasis shelterbelts, together with weakening turbulent exchange inside shelterbelt, and inhibiting vertical transportation of aeolian sediment. Therefore, the quality of aeolian sediment was less inside than outside shelterbelt. The dust is transported under the action of wind, and the sediment flux varies with height during the transport process.

The difference of sand driving wind speed inside and outside shelterbelt was different in wind speed. The frequency of strong sand driving wind exceeding 11 $m{\cdot}s^{-1}$ was decreased significantly inside than that outside shelterbelt in the northeastern edge of Ulan Buh Desert. However, the frequency of sand driving wind between with 5 $m{\cdot}s^{-1}$ - 7 $m{\cdot}s^{-1}$ and between 7 $m{\cdot}s^{-1}$ and 9 $m{\cdot}s^{-1}$ increased inside than outside shelterbelt. When sandstorm passed through the oasis shelterbelts, which acted as a tall rough element, and part of the airflow was lifted up, a relatively high speed free-stream was formed above the shelterbelt canopy. After crossing the shelterbelt, it formed a sinking airflow, which spread in all directions at a certain distance in the leeward zone. Another part of the airflow entered into the shelterbelt. Due to the block and friction of the trees, the airflow consumed a large amount of energy during dispersion, thus a low speed bound-stream was formed under the shelterbelt canopy. Therefore, the frequency of sand driving wind with high speed was decreased, whereas the frequency of sand driving wind with low wind speed was relatively increased inside oasis shelterbelts [34]. Because there was spatial and temporal heterogeneity in the shelterbelt distribution in a specific space with the form of grids [38], our methods on the shelterbelt scale cannot be applied to the regional scale [44]. Therefore, the comprehensive evaluation index system of the environmental benefits of shelterbelts needs to be studied urgently on regional scale. Moreover, it was suggested to focus on deducing the results from shelterbelt scale to multiple shelterbelts or even landscape scales in the future, using interpolation methods to establish models on the time scale [45], and integrating vegetation, soil, climate and other factors, in order to obtain simple and effective results.

5. Conclusion

1. Microclimate of oasis was improved by oasis shelterbelts in the northeastern edge of Ulan Buh Desert, including increasing relative humidity and precipitation, reducing air temperature, surface temperature, evaporation and wind speed inside shelterbelt.

2. During the occurrence of sandstorm, wind speed increased with the increase of height inside and outside shelterbelt, which was expressed as a relationship

of a power function. The wind direction was mainly W, WNW and NE, but the proportion of each direction was different inside and outside shelterbelt.

3. The horizontal aeolian sediment flux decreased significantly with the increase of height outside shelterbelt, while increased slowly with the increase of height inside shelterbelt. The aeolian deposition flux decreased significantly with the increase of height which was expressed as a power function relationship. Wind speed was significantly weakened, the horizontal and vertical fluxes of aeolian sediment flux were reduced when sandstorm passed through oasis shelterbelts.

Acknowledgements

This research was supported by Operation Subsidy of Inner Mongolia Dengkou Desert Ecosystem National Observation Research Station (2021132013), Science and Technology Planning Project of Inner Mongolia Autonomous Region (2020GG0125).

Author details

Fengmin Luo*, Zhiming Xin, Junliang Gao, Yuan Ma, Xing Li and Huaiyuan Liu
Experimental Center of Desert Forestry, Chinese Academy of Forestry, Dengkou, Inner Mongolia, China

*Address all correspondence to: lfm359541965@126.com

References

[1] Wang T. Some issues on oasification study in China. Journal of Desert Research. 2010,30(5):995-998. DOI: http://ir.casnw.net/handle/362004/10812

[2] Hao L, Su X, Singh V P, et al. Suitable oasis and cultivated land scales in arid regions based on ecological health. Ecological Indicators. 2019,102:33-42. DOI: 10.1016/j.ecolind.2019.01.053

[3] Zhang C L, Shen Y P, Li Q J, et al. Sediment grain-size characteristics and relevant correlations to the aeolian environment in China's eastern desert region. Science of the Total Environment. 2018,627:586. DOI: 10.1016/j.scitotenv.2018.01.270

[4] Mu G J, He J X, Lei J Q, et al. A discussion on the transitional zone from oasis to sandy desert: a case study at Cele Oasis. Arid Land Geography. 2013,36(2):195-202. DOI: CNKI:SUN:GHDL.0.2013-02-001

[5] Mctainsh G, Strong C. The role of aeolian dust in ecosystems. Geomorphology. 2007,89(1-2):39-54. DOI: 10.1016/j.geomorph.2006.07.028

[6] Webb N P, Galloza M S, Zobeck T M, et al. Threshold wind velocity dynamics as a driver of aeolian sediment mass flux. Aeolian Research. 2016,20:45-58. DOI: 10.1016/j.aeolia.2015.11.006

[7] Zhou C L, Yang X H, Zhong X J, et al. Dust weather in hinterland of the Taklamakan desert. Arid Zone Research. 2017,34(2):324-329. DOI: 10.13866/j.azr.2017.02.11

[8] Wang T. Desert and desertification in China. Hebei Science and Technology Press.2003. Shijiazhuang

[9] Li Z, Wang N, Li R, Ning K, et al. Indication of millennial-scale moisture changes by the temporal distribution of Holocene calcareous root tubes in the deserts of the Alashan Plateau, Northwest China. Palaeogeography, Palaeoclimatology. Palaeoecology. 2015, 440:496-505. DOI: 10.1016/j.palaeo.2015. 09.023

[10] Zhang C, Li Z, Chen Q, et al. Provenance of aeolian sands in the Ulan Buh Desert, northwestern China, revealed by heavy mineral assemblages. Catena. 2020,193:104624. DOI: 10.1016/j.catena.2020.104624

[11] Wang N A, Li Z L, Cheng H Y, et al. High lake levels on Alxa Plateau during the Late Quaternary. Chinese Science Bulletin. 2011,56:1799-1808. DOI: 10.1007/s11434-011-4498-8

[12] Wang H, Jia X, Li Y, et al. Selective deposition response to aeolian–fluvial sediment supply in the desert braided channel of the upper Yellow River, China. natural Hazards and Earth System Sciences. 2015,15(9):1955-1962. DOI: 10.5194/nhess-15-1955-2015

[13] Li H, Xu F, Li Z, et al. Forest Changes by Precipitation Zones in Northern China after the Three-North Shelterbelt Forest Program in China. Remote Sensing. 2021,13(4):543. DOI: 10.3390/rs13040543

[14] Bao Y F. Windbreak effects of shelterbelt in oases based on wind velocity flow field. Beijing: Beijing Forestry University, 2015.

[15] Van Thuyet D, Van Do T, Sato T, et al. Effects of species and shelterbelt structure on wind speed reduction in shelter. Agroforestry System. 2014,88(2): 237-244. DOI: 10.1007/s10457-013-9671-4

[16] Gao J L, Xin Z M, Liu F, et al. Analysis of the characteristics and affecting factors of atmospheric dustfall in northeastern Ulan Buh Desert. Journal of Arid Land Resources and Environment. 2014,28(8):145-150. DOI: CNKI:SUN:GHZH.0.2014-08-025

[17] Wang J, Dong Z. Research on grain characteristics of falling dust and different surfaces and their relationships on oasis. Journal of Soil and Water Conservation. 2005,19(6):9-11,14. DOI: 10.3321/j.issn:1009-2242.2005.06.003

[18] Luo F M, Xin Z M, Gao J L, et al. Characteristics of near-surface wind speed and sand-dust flux at northeast edge of Ulan Buh Desert. Transactions of the Chinese Society of Agricultural Engineering. 2016,32(24): 147-154. DOI: 10.11975/j.issn.1002-6819.2016.24.019

[19] Reynolds R L, Yount J C, Reheis M, et al. Dust emission from wet and dry playas in the Mojave Desert, USA. Earth Surface Processes and Landforms. 2007,32(12):1811-1827. DOI: 10.1002/esp.1515

[20] Qian Y B, Wu Z N, Zhang L Y, et al. Ground-surface conditions of sand-dust event occurrences in the Southern Junggar Basin of Xinjiang, China. Resources Science. 2006,28(5):185-189. DOI: 10.1016/S1001-0521(06)60029-6

[21] Song Y, Liu L Y, Yan P, et al. Study on the influence of five kinds of underlying matches on sandstorm in North China. Journal of Soil and Water Conservation. 2005,19(6):15-18. DOI: CNKI:SUN:TRQS.0.2005-06-004

[22] Wei Q, Long X, Tian C, et al. Multiscale meteorological characteristics during a sandstorm in Minqin. Arid Zone Research. 2018,35(6):1352-1362. DOI: 10.13866/j.azr.2018.06.12

[23] Zhang H L, Zhang Q L, Ma L Q. Study on microclimate effect of different deploy structure of farmland sheltbelt in Ulanbuh Desert. Journal of Arid Land Resources and Environment. 2009,23(11):191-194. DOI: CNKI: SUN:GHZH.0.2009-11-034

[24] Yang Q, Lei J Q, Wei W S, et al. Effects of artificial oases on climate change trend. Acta Ecologica Sinica. 2004,24(12):2728-2734.

[25] Song J, Wen L, Wang F G, et al. Spatiotemporal dynamics of ecosystem service value in Ulan Buh Desert. Acta Ecologica Sinica. 2021,41(6):2201-2211. DOI: 10.5846/stxb202005141228

[26] Luo F M, Gao J L, Xin Z M, et al. Characteristics of sand-driving wind regime and sediment transport in northeast edge of Ulan Buh Desert. Transactions of the Chinese Society of Agricultural Engineering. 2019,35(4): 145-152. DOI: CNKI:SUN:NYGU.0.2019-04-018

[27] China Meteorological Administration. Ground Meteorological Observation Specification. Beijing: Meteorological Press, 2003: 11. ISBN: 7-5029-3690-4

[28] Zhui R Z, Tan G R, Wang S L. Introduction to Applied Climatology. Beijing: Meteorological Press, 2005: 2-3. ISBN: 7-5029-4036-7

[29] Dong Z B, Zhao A G, Deng J. The surveyor of horizontal sand-dust flux: China, ZL0313453414. 2008.

[30] Zhang F, Tashpolat·Tiyip, Ding J L, et al. The change of land use/cover and characteristics of landscape pattern in arid areas oasis: a case study of Jinghe County, Xinjiang Province. Acta Ecologica Sinica. 2009,29(3):1251-1263. DOI: 10.3321/j.issn:1000-0933.2009.03.021

[31] Zhang P, Zhao X F, Zhang T, et al. Leaching salinity effect of drip irrigation on soil of oasis shelter forests in arid area. Transactions of the Chinese Society of Agricultural Engineering. 2011,27(5):25-30. DOI: 10.3969/j.issn.1002-6819.2011.05.005

[32] Zhang L Y, Wang R A. Deserts and Oasis in China. Lanzhou: Gansu Education Press, 1994. ISBN:7-5423-0477-1

[33] Fang L D, Ning Q R, Guo J J, et al. Hydraulic limitation underlies the

dieback of *Populus pseudo-simonii* trees in water-limited areas of northern China. Forest Ecology and Management. 2020,483:118764. DOI: 10.1016/j.foreco. 2020.118764

[34] Kong D S, Jin B W, Jin M, et al. Microclimate effects of farmland shelterbelt in middle reaches of Heihe basin. Journal of Arid Land Resources and Environment. 2014,28(1):32-36. DOI: 10.3969/j.issn.1674-6813.2014.04.097

[35] Wang X L, Wu Y J. Study on the microclimate effect of forest shelter network in wetland agricultural ecosystem of Jianghan plain-lake district. Resources and Environment in the Yangtze Basin. 1999,8(4):418-420. DOI: CNKI:SUN:CJLY.0.1999-04-013

[36] Yang Y P. Research on Shelterbelt of Upper Yangtze River (Chuanjiang). Beijing: Science Press, 1993: 7-31. ISBN: 7-03-004164-X

[37] Zhao M, Zhan K J, Yang Z H, et al. Characteristics of the lower layer of sandstorms in the Minqin desert-oasis zone. Science China: Earth Sciences. 2011,41(2):234-242. DOI: 10.3724/ SP.J.1011.2011.00181

[38] Liu X D. Analyses of the spring dust storm frequency of Northern China in relation to antecedent and concurrent wind, precipitation, vegetation, and soil moisture conditions. Journal of Geophysical Research. 2004,109(16): 1-16. DOI: 10.1029/2004JD004615

[39] Deng X Y, Yao J Q, Liu Z H. Spatiotemporal dynamic change of vegetation coverage in arid regions in Central Asia based on GIMMS-NDVI. Arid Zone Research. 2017,34(1):10-19. DOI: 10.13866/j.azr.2017.01.02

[40] Zhang Z C, Dong Z B, Zhao A G, et al. Features of sand drift movement in desert: A case study at sand drift observation station of Chinese Academy of Sciences. Arid Zone Research. 2007,24(4):550-555. DOI: 10.1016/ S1872-5791(07)60044-X

[41] Yang X H, Li H J, He Q, et al. Blown sand activities in spring in the desert transitional zone of the Taklimakan Desert: A case in Xiaotang area. Journal of Desert Research. 2012,32(4):915-920. DOI: CNKI:SUN:ZGSS.0.2012-04-004

[42] Yang D L, Wang X Q, Hu Y F, et al. Vertical distribution of aeolian sand mass flux based on field observation at southern rim of the Taklimakan Desert. Journal of Desert Research. 2012,32 (3):631-639. DOI: CNKI:SUN:ZGSS.0. 2012-03-009

[43] Yang X H, He Q, Ali Mamtimin, et al. A field experiment on dust emission by wind erosion in the Taklimakan Desert. Acta Meteorologica Sinica. 2012,26(2):241-249. DOI: 10.1007/s13351-012-0209-x

[44] Waring R H, Running S W. Forest ecosystem analysis at multiple time and space scales. Forest Ecosystems. Amsterdam: Elsevier, 2007:1-16. ISBN: 9780123706058

[45] Deng R X, Zhang S W, Li Y. Protective effects of farmland shelterbelts in Northeast China: a field scale analysis. Chinese Journal of Ecology. 2009,28(9):1756-1762. DOI: CNKI:SUN:STXZ.0.2009-09-015

Section 2

Desert Vegetation

Chapter 3

Remote Sensing Based Quantification of Forest Cover Change in Somalia for the Period 2000 to 2019

Sylus Kipngeno Musei, Justine Muhoro Nyaga and Abdi Zeila Dubow

Abstract

Deforestation is a driver of land degradation and a major environmental problem in Somalia, and has been linked to frequent incidences of drought over the years. Monitoring of changes in forest cover is therefore critical for the country's environment. The problem of land degradation has been worsened by the large scale charcoal production that is witnessed in the country. This study aimed at estimating forest cover change between 2000 and 2019 in Somalia using Landsat-based forest cover datasets. Google Earth Engine (GEE), a cloud based computing system was used to provide a platform for this analysis. Based on the 30% threshold recommended by International Geosphere Biosphere Program for differentiating forest from non-forest trees, approximately 23% forest cover loss was found, from 87, 294 hectares in 2000 to 67, 199 hectares in 2019. Most of the country's forest is within the southern and central parts of the country, and significant forest cover losses occurred mainly around Mogadishu and Kismayo port throughout the study period. There is therefore a need for the Federal Ministry of Environment and environment ministries in the federal member states to design mechanisms and strategies for restoration of the degraded forests and to curb deforestation.

Keywords: remote sensing, land degradation, Somalia, Google earth engine, charcoal production

1. Introduction

Forests form a critical component of global carbon cycle [1] in form of woody above ground biomass. Their destruction through deforestation disrupts this cycle leading to serious negative consequences on ecosystem functioning. Deforestation may be caused by both natural and anthropogenic factors. Anthropogenic factors included population growth, technological advancements and cultural norms, and these may accelerate deforestation at both local and regional levels [2, 3]. This in turn threatens carbon storage, watershed protection and biodiversity particularly in developing countries [4, 5] where they have been rampant. Forest cover is the most important indicator of forest degradation. The 2015 forest resource assessment

(FRA) report of the Food and Agricultural Organization (FAO) indicated that the global forest cover was about 4 billion hectares [6] (about 30.6% of world's land area). Tropical forests covered about 20% of the global land area which dropped to less than 7% at the end of the 20th century [1]. The tropical region experienced the greatest loss and acquisition of forests among the four climate domains of the world, and the highest loss ratio (3.6 to >50% tree cover), indicating the prevalence of the dynamics of deforestation [4].

Conservation of forests improves livelihoods of communities dependent on them for various goods and services. Globally, over 1.6 billion people rely on forests as a source of food, fuel, medicines, water and for cultural use either directly or indirectly [7]. All over the world, forests play an important role in minimizing greenhouse gas emissions [8] and removing excess carbon dioxide from the atmosphere through carbon sequestration. In sub-Saharan Africa, at least 80% of the urban population and 90% rural population depend on charcoal as a source of energy for cooking and warming houses [9, 10]. The main source of energy for cooking in Somalia is charcoal [11–13] which is obtained from forests through deforestation. Furthermore, Somalia being a pastoralist country depends on trees and shrubs for livestock feeds during dry seasons [14, 15]. The resultant forest destruction could lead to devastating impacts on the natural resources which are highly depended on by the pastoral communities [16]. Foreign demand for charcoal has also escalated deforestation rates in Somalia with 4.4 million trees cut down annually for making of export charcoal [13]. As a major export product, charcoal is the major source of income for 70% of the poor and middle-income pastoralists since it requires small capital for production [17–19]. This has significantly contributed to destruction of the country's natural forest resources [20].

In addition to the basic conservation concern about deforestation in Somalia, the international "Reducing Emissions from Deforestation and forest Degradation" (REDD+) initiative has recently become a major component of forest conservation in the country. This is a global effort aimed at conserving, sustainably managing, enhancing, and monitoring forests [21] and thus requires the government through the forestry department to account for the current forest extents for sustainable forest management. This may further require quantification of greenhouse gas emissions and reductions from forests. These therefore necessities quantification of forest cover changes across the country. However, quantifying forest cover changes through field surveys is limited by time and resource availability.

Remotely sensed data however provides alternative means for monitoring forest cover changes [22, 23]. Temporal and spatial components of observational remotely sensed data have improved the quality and the quantity of satellite data by enhancing their applicability in monitoring environmental changes. A number of satellite-derived products have been developed with a focus on global forest cover. MODIS images that have been captured by Terra and Aqua satellites since 2000 have enabled formation of the first annual forest cover product, MODIS Vegetation Continuous Field (VCF) [24, 25]. This product is limited by the patchiness of tree cover changes which have smaller spatial sizes compared to the MODIS VCF [26]. This deficiency led to the production of Landsat based global continuous field cover product (30 m) [27]. The Landsat based product has finer spatial resolution compared to the MODIS based VCF and thus facilitates a more accurate forest cover change assessment. Besides the finer spatial resolution, Landsat data also have high amounts of cloud cover contamination as well as infrequent revisit coverage [28]. Due to this limitation global mosaics were produced for the years 1975, 1990, 2000, 2005 and 2010 [29]. However, more recently a global 30 m forest cover change product was published during 2000–2012 [4]. A latest version of this product is available for the period between 2000 and 2019.

Even with the increased availability of high-resolution satellite data, it has always been difficult analyzing high spatial resolution satellite data, for example, at a country level. This is because at such a large scale, huge volumes of data are required to cover the entire country [24]. Google Earth Engine (GEE), a cloud-based platform however overcomes this challenge since it allows for processing of huge geospatial datasets [29]. Satellite resources have been used to monitor forest cover changes in various countries across the globe [4]. In the case of Somalia, there have been a few applications but none have quantified forest cover change across the entire country. In this study we utilize the historical earth observation datasets to map and quantify forest cover over time in Somalia by utilizing the capabilities of Google Earth Engine (GEE). The main objective was to evaluate the spatial and temporal patterns of forest cover from 2000 to 2019 across Somalia.

2. Materials and methods

2.1 Study area

This assessment covered the whole of Somalia (**Figure 1**). The country extends from the equator to the north of Gulf of Aden. It is one of the Horn of African countries and borders Kenya and Ethiopia to the west, Indian Ocean to the east and Djibouti to the northwest. The country's population was estimated at 14.74 million in 2017 by the World Bank. The population increased from 8,872,254 in 2000 to 15,442,905 in 2019 (**Figure 2**). This is equivalent to an increase of 6,570,651 (74.06%).

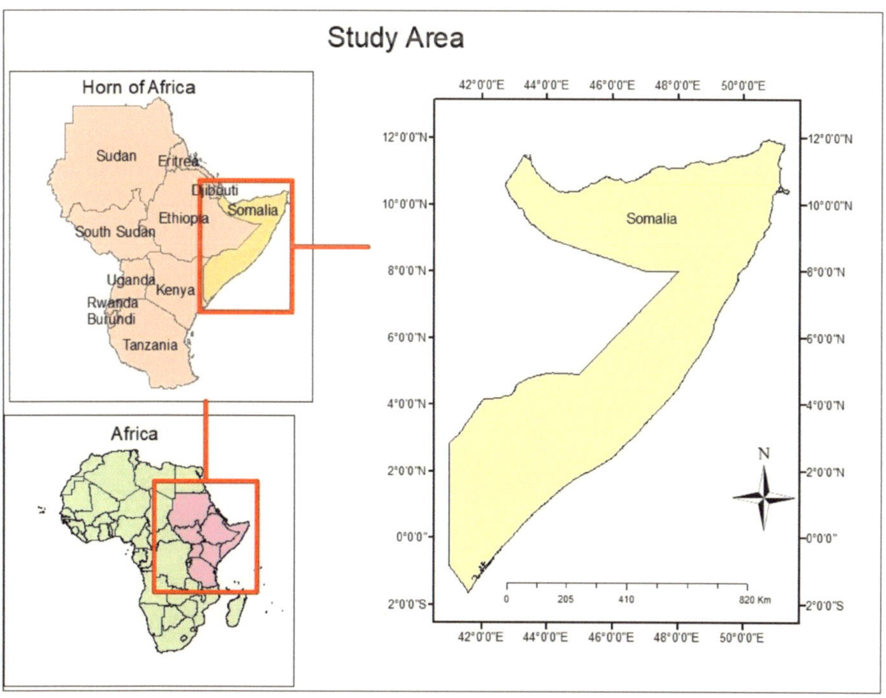

Figure 1.
Map of the study area. Somalia is part of the horn of Africa and the eastern-most country in the African continent.

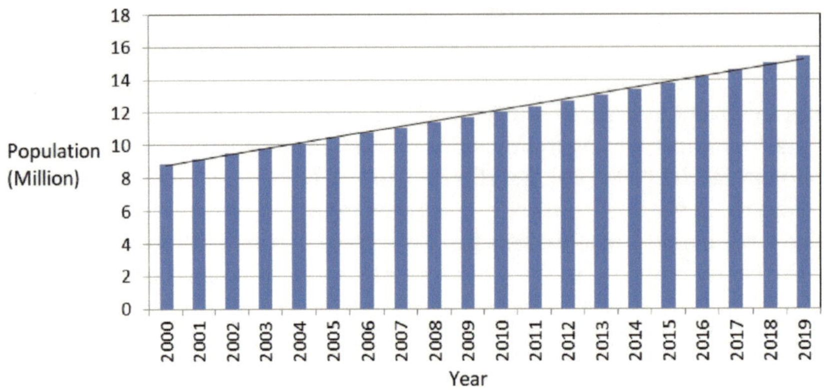

Figure 2.
Population variations between 2000 and 2019 (source: The World Bank database).

Somalia covers a total land area of 637,660 km^2 most of which may be classified as either arid or semi arid. Its administrative and commercial capital is Mogadishu. According to [30], at least 50% of the country's economy is dependent on pastoralism.

Topographically, the country is divided into several physiographic zones. The northern part of the country receives high annual temperatures and forms the driest zone. Plateaus cover the central and southern parts bordering the coastline. The two main perennial rivers (Webe Shebele and Jubba), whose source is Ethiopian plateau, are found in southern Somalia. Between these two rivers is a highly productive plain which is occasionally affected by floods. Somalia's coastline covers 3,025 kilometers and is the longest in Africa.

Generally, the country is hot throughout the year with an annual average day temperature of 27°C. This is attributed to country's close proximity to the equator and the low rainfall amounts received. The northern parts experience highest temperatures of 45°C around Karkaar Mountain in summer while the lowest temperatures are experienced to the south of Somalia along the coastline largely due to the effect of the sea breeze. Somalia received little amount of rainfall throughout the study period with an average of 20.96 mm per year. The lowest (18.5 mm) and highest (25.6 mm) amount of rainfall was received in 2010 and 2019 respectively. Although there is a generally upward trend in the amounts of rainfall received over the years from 2000 to 2019, the total annual amount of rainfall is still small as shown in **Figure 3**.

Somalia has two climate zones, arid and semiarid zone. The semiarid zone covers the northern and southwest parts of the country and the arid zone covers the central and the northernmost parts [31]. The semiarid zone receives medium rainfall and is suitable for rain fed agriculture whereas the arid zones receiving low rainfall are preferably used for pastoralism. The most common tree species is *Acacia commiphora* [32] which covers about 50% of southern Somalia. Others include: Apple ring acacia (*Acacia albida*), Egyptian thorn (*Acacia nilotica*), Gum arabic (*Acacia senegal*), Umbrella thorn (*Acacia tortilis*), soapberry tree (*Balanites aegyptiaca*), Myrrh tree (*Commiphora myrrah*), common tug tree (*Conocarpus lancifolius*), Yihib nut tree (*Cordeauxia edulis*), Spiny desert tree (*Terminalia spinosa*), Pencil cedar (*Juniperus excels*), Tamarisk (*Tamaria aphylla*) and Tamarind (*Tamarindus indica*).

2.2 Dataset and the analysis platform

This study adopted the word wide forest data that was developed by Hansen [4]. The dataset was created from the annual 30 m resolution Landsat data. The

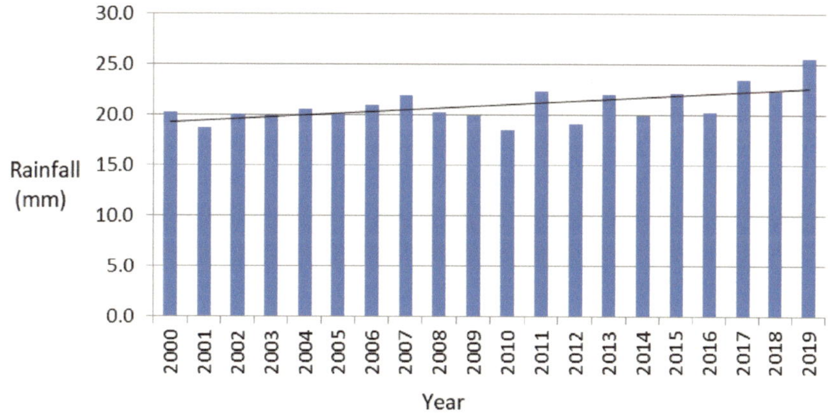

Figure 3.
Annual rainfall trends from 2000 to 2019 (source: The World Bank database).

products being global forest cover at year 2000 and annual forest cover loss and gain. In this dataset, vegetation above 5 m in height is defined as trees. Trees are expressed as '2000 Percent Tree Cover' percentage per output plot cell. Forest Cover Loss is defined as a change from forest to non-forest status in the period 2000–2019. Forest gain is defined as the opposite of loss, that is, the change in non-forest to forest in the period 2000–2019.

Google Earth Engine (GEE) enables cloud computing processing of Landsat images through its online system. This helps to navigate the challenge of processing Landsat images at global scale which is normally expensive in terms of time and resources [4]. Landsat images are ideal for monitoring environmental changes at local scale because of the 30 m spatial resolution [33]. Since 1972, Landsat have been used for land cover change analysis [34–37]. Due to its ability to analyze large volumes of remotely sensed datasets, GEE was preferably selected for Landsat data analysis in this study. For ease and fast workflow, a web-based JavaScript API was developed in the GEE Code Editor Feature.

2.3 Data analysis

First, forest cover in the starting year, 2000, was analyzed using Google Earth Engine. This was then followed by evaluation of temporal patterns of forest cover losses from the year 2001 to 2019. GEE enabled selection of the whole Somalia from the global Hansen [4] forest dataset using Google Earth Engine code editor with JAVA script API.

3. Results

3.1 Spatial trends of forest cover in Somalia

This internet link (https://code.earthengine.google.com/6d67bcdf27fd3139f89c0a1af94a3c9d) leads to the script developed for this analysis. The area covered by forest in Somalia in the year 2000 was 8,729,400 ha. This is approximately 13% of the total land area in the country. **Figure 4** shows the extent of forest cover in the year 2000. Forest mainly covers the southern part of the country with small patches to the north. The largest part of the country is dry as shown in the **Figure 4** with the green patches showing the forested areas and the non-green parts shows non-forested areas.

3.2 Forest cover loss

The spatial distributions of forest cover loss throughout the country are shown in **Figure 5**. The red spots show areas where forest cover has been lost. From this figure, it's clear that areas around the country's capital (Mogadishu) such as Jowhaar, Mahaday, Wyne and Merca have suffered the highest forest cover loss over the years. Apart from the capital, Kismayo and the surrounding areas of Dujuma, Haramka, Jamaame, Jilib and Yoontoy also experienced high forest cover loss throughout the study period.

3.3 Temporal variations of forest cover loss

The patterns of forest loss were computed based on Hansen's [4] forest cover change dataset. Temporal changes in forest loss areas were calculated from 2000 to 2019 also based on Hansen's [4] forest cover change data. **Figure 6** shows the

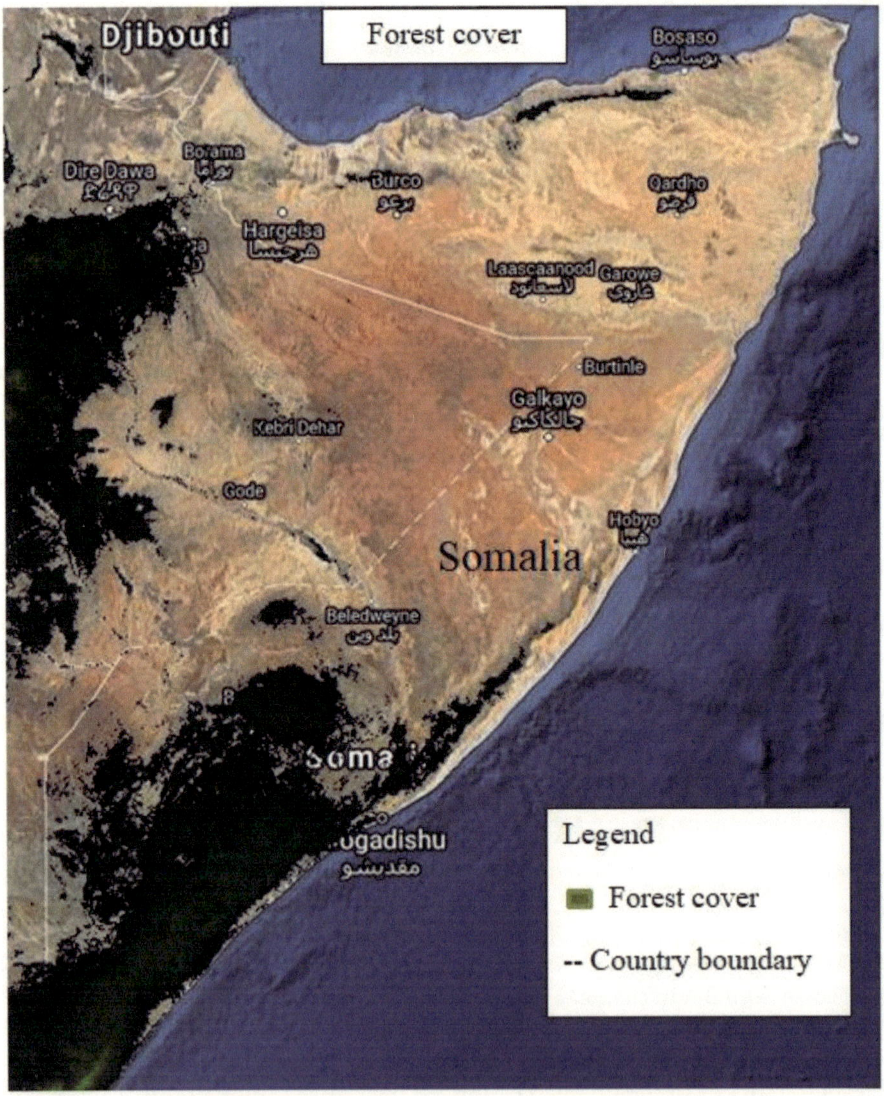

Figure 4.
Forest cover in 2000.

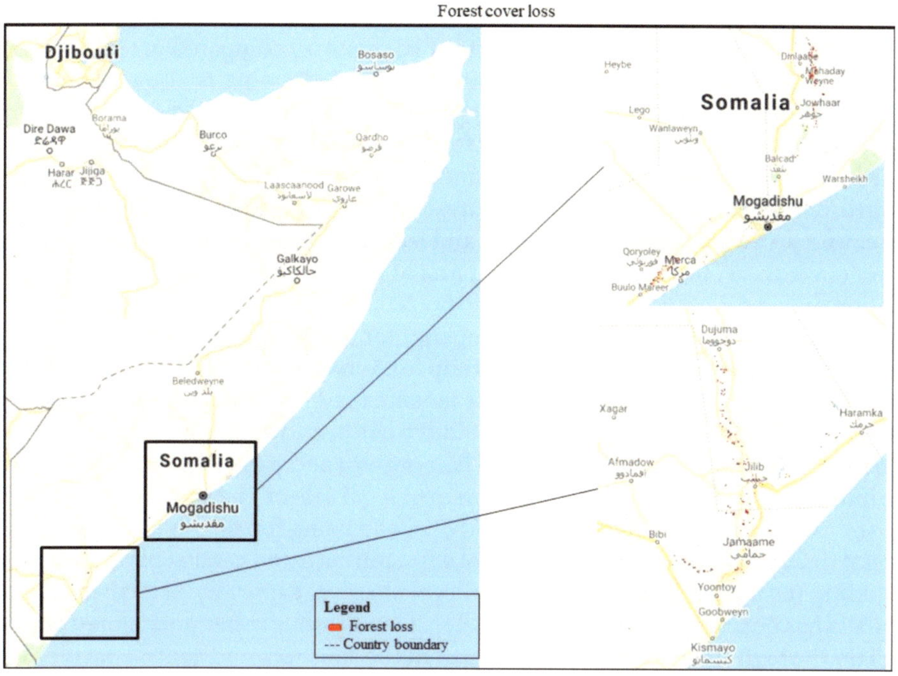

Figure 5.
Forest cover loss between 2000 and 2019. The red spots show areas of forest loss throughout the study period.

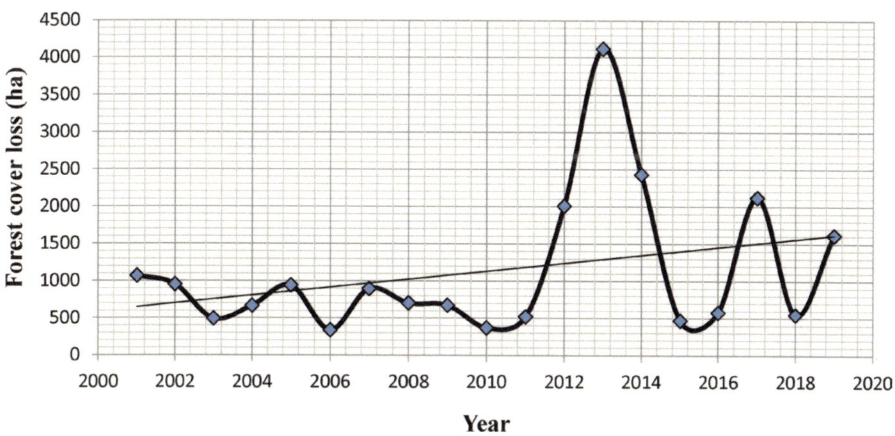

Figure 6.
Annual trends of forest area loss from 2000 to 2019.

yearly trend of forest cover loss in Somalia throughout the study period. There is generally upward trend of forest loss with the peak occurring in the year 2013 when 4116.76 ha of forest cover were lost. The least forest cover loss happened in the year 2006 when 336.72 ha of forest was lost.

4. Discussions

In this study, the year 2000 was used as the baseline, and during that year, most of the forest in Somalia was in the southern parts with only small patches of forests in the northern parts of the country (**Figure 4**). This pattern of forest cover may be explained by the significant variations in climatic conditions between the northern and southern parts of the country. Relatively high temperatures are experienced throughout the year in the northern parts of the country compared to the southern parts that have relatively lower temperatures. The southern parts also receive moderate amounts of rainfall on occasional basis and these have been shown to support vegetation growth [38], thus the higher forest cover relative to the north.

Forest cover loss was unevenly distributed throughout the country as shown in **Figure 5**. It is evident that most of the forest loss occurred to the southern parts of the country, mainly around Mogadishu and Kismayo, probably due to the higher forest cover there. Besides, it is in these parts of the country where charcoal production and use is very high. Research on household sources of energy in Somalia has established that the main source of energy for Mogadishu is charcoal [11]. In 2011, the United Nations Monitoring Group for Somalia and Eritrea reported that charcoal was also an important source of income for Al Shabaab [39]. The proliferation of this armed militia group in the country has been a source of insecurity and underdevelopment from 2008 to date. Their growth and expansion may be partly attributed to the incomes generated from the illegal sale of charcoal produced in the southern parts of Somalia. Besides Al Shabaab being forced out of the country's capital in 2011, they continued to control vast southern and central parts of Somalia including the port of Kismayo. Most of illegal charcoal export by Al Shabaab from Somalia has been through this port [40, 41]. These among other unexplored reasons explain the high levels of forest destruction in the southern and central parts of Somalia throughout the study period. Some of these reasons may include inappropriate land use and extraction of other natural resources besides the forests. These practices have been so rampant in Somalia that forest cover loss in this country has also significantly contributed to forest cover loss across the African continent. It has been shown that Somalia is responsible for 6% of all trees lost in the continent with an estimated annual forest cover loss of 76,757 hectares [6].

Results from this study show a general upward increasing trend of forest cover loss throughout the study period (**Figure 6**). The peak of forest cover loss was in 2013 when over 41,00 ha of forest area was lost. This can be attributed to the high charcoal production (approximately 24, 000 metric tons of charcoal produced between 2011 and 2013 in southern parts of Somalia) and sale that was experienced during this period in Somalia [42]. During this time, the charcoal industry was dominated by the Al Shabaab militant group which controlled the region at the time, and had no regard for environmental conservation. Their main goal was to raise money to finance their terrorist activities, and charcoal production and sale was a one of their major sources of income.

Apart from charcoal production, the high losses in forest cover can also be attributed to the high rates of population increase in the Somalia (**Figure 2**). Population increased exponentially throughout the study period, with a total percentage increase of 75% between the year 2000 and 2019. Increase in population exerts more pressure on economic land resources consequently resulting in their degradation and eventual destruction. The gradual increase in population in Somalia over the study period exerted considerable pressure on the forest resources which are major land economic resources in the country. More trees were cut down to provide energy for cooking as firewood or charcoal, and also for construction timber among other forest products. Another thing that may explain the high forest cover losses in Somalia is the low annual precipitation the country received over the study period. Although the annual rainfall had a generally increasing trend between 2000 and 2019, the average annual rainfall for the entire period remained low at approximately 21 mm. Low rainfall slows down regeneration of degraded forests and may lead to drying out of the young seedlings planted to replace the cut down trees.

5. Conclusions

From this study, the following conclusions can be made;

1. There is an increasing upward trend of forest cover loss throughout the study period in Somalia with the peak loss recorded in 2013.

2. Most forests cover the central southern parts of Somalia. Forest cover loss mainly occurred in areas around Mogadishu and Kismayo port.

3. Google Earth Engine combined with satellite data provides a quick means to assessing forest cover changes at regional and global scale thus saving on time and money resources. It also makes it easy to study environmental resources in Somalia besides the high insecurity in the country.

These results informs the need for the department of Forestry under the Federal ministry of Environment to design policies and targeted actions towards restoration of degraded forests and protection of the non-degraded forests in Somalia. Further research can be done to ascertain the causes of the upward trend of forest cover loss, especially after Al shabaab militants were driven out of their bases in Southern Somalia. This information would inform government actions geared towards management of forest resources in the country.

6. Recommendations

To increase forest cover and enhance management of forest resources, the national government and the federal member states of Somalia, through the respective ministries of environment and forestry need to consider the following recommendations.

1. Develop a long lasting forest cover change monitoring platform to monitor dynamics of forest cover for informed and timely protection efforts based on the magnitude of destruction.

2. Develop a national afforestation and reforestation program and framework to guide country's efforts towards restoration of degraded forests. For example, fast growing trees can be planted in areas that have experienced vast forest destruction such as Kismayo and the country's capital, Mogadishu. Some of the fast growing desert trees that can be planted as a way of mitigating deforestation impacts include: Chitalpa tree (*Chitalpa x tashkentensis*), Chinese elm (*Ulmus parvifolia*), California pepper tree (*Schinus molle*) and *Melia volkensii* [43].

3. Create public awareness on the risks and implications of forest destruction so that the community sees the need to protect forests and move away from activities such as charcoal production which is the leading driver of deforestation in the country.

4. Create and strengthen existing avenues for community involvement in forest management. Such may include establishment of Community Forest Associations which could co-manage the forest resources with government agencies. This will enhance cooperation among various stakeholders in the forestry sector. Community Forest Association are important since they create a sense of possession among community members and this can go a long way in protection of the forests.

5. The government through the ministry of education should include forest protection and management topics in the schools curricula so that a culture of protection of forest resources is inculcated to learners at an early age. They could also develop and include risk and vulnerability management programs that are related to forestry at all levels in educational sector.

Author details

Sylus Kipngeno Musei[1], Justine Muhoro Nyaga[1*] and Abdi Zeila Dubow[2]

1 Department of Biological Sciences, University of Embu, Embu, Kenya

2 Department of Land and Water Management, University of Embu, Embu, Kenya

*Address all correspondence to: nyagajm@gmail.com

References

[1] Encyclopaedia Britannica (2018). Status of the world's tropical forests, available on the web: https://www.britannica.com/topic/Status-of-the-Worlds-Tropical-Forests-1673338

[2] Geist, H.J., Lambin, E.F. (2002). Proximate causes and underlying driving forces of tropical deforestation tropical forests are disappearing as the result of many pressures, both local and regional, acting in various combinations in different geographical locations. BioScience 52, 143-150.

[3] Hosonuma, N., Herold, M., Sy, V.D., Fries, R.S.D., Brockhaus, M., Verchot, L., Angelsen, Arild, Romijn, E. (2012). An assessment of deforestation and forest degradation drivers in developing countries. Environmental Research Letters. 7, 044009.

[4] Hansen, M. C., Potapov, P. V., Moore, R., M., Hancher, S. a. Turubanova, a., Tyukavina, D., Thau, S. V., Stehman, S. J., Goetz, T. R. Loveland, a., Kommareddy, a., Egorov, L., Chini, C. O., justice, J., Townshend R. G. (2013). High-resolution global maps of 21st-century forest cover change. Journal of Science, 342, 850-853.

[5] Tyukavina, A., Baccini, A., Hansen, M.C., Potapov, P.V., Stehman, S.V., Houghton, R.A., Krylov, A.M., Turubanova, S., Goetz, S.J. (2015). Aboveground carbon loss in natura land managed tropical forests from 2000 to 2012. Environmental Research Letters. 10, 074002.

[6] FRA (2015). The global forest resources assessment, Rome, FAO (Food and Agriculture Organization of the United Nations).

[7] Zulu, L.C. and Richardson, R.B. (2013) Charcoal, livelihoods, and poverty reduction: Evidence from sub-Saharan Africa. Energy for Sustainable Development, 17, 127-137.

[8] Bailis, R., Ezzati, M. and Kammen, D.M. (2005) Mortality and greenhouse gas impacts of biomass and petroleum energy futures in Africa. Science, 308, 98-103.

[9] Brink, A.B., Eva, H.D. and Bodart, C. (2012). Is Africa losing its natural vegetation?—Monitoring trajectories of land-cover change using landsat imagery. In: Giri, C.P., Ed., Remote Sensing of Land Use and Land Cover: Principles and Applications, CRC Press, Taylor and Francis, Boca Raton, 369.

[10] Sedano, F., Silva, J.A., Machoco, R., Meque, C., Sitoe, A., Ribeiro, N., Tucker, C., et al. (2016). The impact of charcoal production on forest degradation: A case study in Tete, Mozambique. Environmental research letters, 11, Article ID: 094020.

[11] Robinson, A. (1988) Charcoal-making in Somalia: A look at the bay method. An international journal of the forestry and Food Industries, 159.

[12] IPCC (2014) Africa. In: Climate Change 2014: Impacts, adaptation, and vulnerability. Part B: Regional Aspects. Contribution of working group ii to the fifth assessment report of the intergovernmental panel on climate change, Cambridge University Press, Cambridge, and New York.

[13] SWALIM (2014) Detection of charcoal production sites on southern somalia using very high-resolution imagery.

[14] Little, P. D. (2005). Pastoralism in a stateless environment: The case of the southern Somalia borderlands. In Geography Research Forum, 25, 128-147.

[15] IUCN (2006) Country environmental profile for Somalia.

[16] USAID (2014) Environmental and natural resource management assessment.

[17] UNEP (2005) The state of the environment in Somalia: A Desk study.

[18] Beier, A.C. and Stephansson, E. (2012) Environmental and Climate Change Policy Brief Somalia.

[19] WB (2016) Data. The World Bank.

[20] Rembold, F., Oduori, S.M., Gadainb, H. and Tose, P. (2013) Mapping charcoal driven forest degradation during the main period of Al Shabaab control in southern Somalia. Energy for Sustainable Development, 17, 510-514.

[21] UN-REDD (2015) UN-REDD Programme strategic framework 2016-20.

[22] Rogan, J and Chen, D M. (2004), Remote sensing technology for mapping and monitoring land-cover and land-use change, Progress in Planning 61, 301-325.

[23] Belgiu, M and Drăgut, L. (2016)., Random forest in remote sensing: A review of applications and future directions., ISPRS Journal of Photogrammetry and Remote Sensing 114, 24-31.

[24] Hansen, M.C., DeFries, R.S., Townshend, J.R.G., Carroll, M., Dimiceli, C., Sohlberg, R.A. (2003). Global percent tree cover at a spatial resolution of 500 meters: First results of the MODIS vegetation continuous fields algorithm. Earth Interactions. 7, 1-15.

[25] DiMiceli, C., Carroll, M., Sohlberg, R., Huang, C., Hansen, M., Townshend, J. (2011). Annual Global Automated Modis Vegetation Continuous Fields (mod44b) at 250 M Spatial Resolution for Data Years Beginning Day 65, 2000-2010, Collection 5 Percent Tree Cover. USA: University of Maryland, College Park, MD.

[26] Jin, S.M., Sader, S.A. (2005a). Comparison of time series tasseled cap wetness and the normalized difference moisture index in detecting forest disturbances. Remote Sensing and Environmental monitoring. 94, 364-372.

[27] Sexton, J.O., Song, X.-P., Feng, M., Noojipady, P., An and, A., Huang, C.Q. (2013). Global, 30-m resolution continuous fields of tree cover: Landsat-based rescaling of MODIS vegetation continuous fields with lidar-based estimates of error. International Journal of Digital Earth. 6, 427-448.

[28] Townshend, J.R., Masek, J.G., Huang, C.Q., Vermote, E.F., Gao, F., Channan, S., et al. (2012). Global characterization and monitoring of forest cover using Landsat data: Opportunities and challenges. International Journal of Digital Earth. 5, 373-397.

[29] Hansen, M.C., Stehman, S.V., Potapov, P.V., Arunarwati, B., Stolle, F., Pittman, K. (2009). Quantifying changes in the rates of forest clearing in Indonesia from 1990 to 2005 using remotely sensed data sets. Environmental Research Letters. 4, 1-12.

[30] Alwesabi, M. (2012). MODIS NDVI Satellite Data for Assessing Drought in Somalia during the Period 2000-2011. Student thesis series INES.

[31] Menkhaus, K. (2014). Calm between the storms? Patterns of political violence in Somalia, 1950-1980. Journal of Eastern African Studies. 8(4), 558-572.

[32] Hartley, A. J., Nelson, A., Mayaux, P., and Grégoire, J. M. (2000). The assessment of African protected areas. Journal of Biogeography, 31, 861-877.

[33] Woodcock, C.E., Allen, R., Anderson, M., Belward, A.,

Bindschadler, R., Cohen, W., Gao, F., Goward, S.N., Helder, D., Helmer, E., Nemani, R., Oreopoulos, L., Schott, J., Thenkabail, P.S., Vermote, E.F., Vogelmann, J., Wulder, M.A., Wynne, R. (2008). Free access to landsat imagery. Science, 320, 298-201.

[34] Alberti, M., Weeks, R., Coe, S. (2004). Urban land cover change analysis in central puget sound. Remote Sensing. 70, No. 9, 1043-1052.

[35] Yuan, F., Sawaya, K.E., Loeffelholz, B.C., Bauer, M.E. (2005). Land cover classificationand change analysis of the twin cities (Minnesota) metropolitan area by Mul-titemporal landsat remote sensing. Remote Sensing and Environmental Monitoring. 98 (2-3), 317-328.

[36] Rawashdeh, S.A., Saleh, B. (2006). Satellite monitoring of urban spatial growth inamman area, Jordan Journal of Urban Planning. 132, No. 4, 211.

[37] Bagan, H., Yamagata, Y. (2012). Landsat analysis of urban growth: How Tokyo becamethe world's largest megacity during the last 40 years. Remote Sensin and Environmental Monitoring. 127, 210-222.

[38] Hadden, R.L. (2007). "The Geology of Somalia: A Selected Bibliography of Somalian Geology, Geography and Earth Science." Engineer Research and Development Laboratories, Topographic Engineering Center

[39] Chopra, J. (2014). Report of the Monitoring Group on Somalia and Eritrea Pursuant to Security Council Resolution 2111 (2013): Somalia.

[40] Rembold, F., Oduori, S. M., Gadain, H., and Toselli, P. (2013). Mapping charcoal driven forest degradation during the main period of Al Shabaab control in southern Somalia. Energy for Sustainable Development, 17(5), 510-514.

[41] Bolognesi M., Leonardi U., Vrieling A., Rembold F., Gadain H. (2014). Detection of Charcoal Production Sites in Southern Somalia Using Very High Resolution Imagery. Technical Project Report. FAOSWALIM, Nairobi, Kenya.

[42] Bolognesi, M., Vrieling, A., Rembold, F., and Gadain, H. (2015). Rapid mapping and impact estimation of illegal charcoal production in southern Somalia based on WorldView-1 imagery. Energy for Sustainable Development, 25, 40-49.

[43] Muok, B., Mwamburi, A., Kyalo, E., and Auka, S. (2010). Growing Melia volkensii. A Guide for Farmers and Tree Growers in the Drylands, 3. Nairobi.

Chapter 4

Cultivation of Artificial Algal Crust and Its Effect on Soil Improvement in Sandy Area

Jinchao Feng, Wei Li and Linna Ma

Abstract

Algae are the pioneer species of biological soil crusts. Cyanobacteria, microschwannophyta and pseudocladophyta can form fixed quicksand algae crusts on the surface of sand surface. Through artificial culture, soil crusts can be formed in a short time. The development and succession of algeal-sand crust promoted the enrichment of nutrients in the sand surface layer, and created conditions for the reproduction of micro-soil organisms and the colonization of herbaceous plants, thus promoting the desert ecosystem to enter a virtuous cycle. This chapter will focus on the cultivation process of artificial soil crust and its effect on soil improvement (soil organic matter and nitrogen) in sandy areas. In conclusion, the application of algal solution can rapidly form algal crusts, and according to the research results, the formation of algal crusts can significantly improve the chemical and biological properties of soil.

Keywords: algal crust, soil organic carbon, microbial carbon and nitrogen

1. Introduction

China is one of the countries most seriously affected by desertification in the world. Desertification not only causes the imbalance of ecosystem, but also reduces the area of arable land, and brings serious impact and harm to industrial and agricultural production and people's life. The arid desert and desertified land in northwest China have become one of the main sources of sandstorms in China and even in the Asia-Pacific region, causing great losses to the country, society and economy [1]. Therefore, desertification land management is the urgent need of the country in ecological construction and environmental protection. For a long time, afforestation and grass planting have been the main ways to control the desert, and some results have been achieved in practice. However, desertification control through traditional methods such as afforestation can sometimes be difficult to achieve, for example, in areas with less than 200 mm of rainfall. Therefore, it is necessary to have new ideas to control desertification [2].

With the increase of the construction history and economic investment of artificial vegetation in sandy land in China, the research on the development mechanism of sandy soil under artificial vegetation and the effect of soil modification by plants have been gradually strengthened [1]. Especially in recent years, the research on the ecological role of Biological Soil Crusts (BSC) has attracted great attention after the mobile sandy land was fixed. Sand surface after intervention algae and algae growth through its metabolism, driving the growth of soil heterotrophic microorganisms, increase the biodiversity in the desert surface, thus promotes the mineralization process of sand and soil weight circulation and flow, which is beneficial to improve soil physical and chemical properties, and increase soil organic matter, total nitrogen and total phosphorus content [1].

Biological soil crust refers to the complex surface cover formed by the interaction of cyanobacteria, green algae, lichens, mosses, microorganisms and other related organisms on and below the soil surface with soil surface particles through mycelium, pseudoroots and secretions. As ground cover, it generally exists in arid and semi-arid regions of the world, and its coverage accounts for 70% of bare land area [3]. It is a component of dryland ecosystem and an important landscape feature [4, 5].

The existence of biological crust plays a crucial ecological role, for the physical character, it could improve soil pore structure, reduce soil bulk density [6], reduce rainfall infiltration rate [7, 8], effectively ease the rain splash erosion and rainfall runoff scouring effect [6], add sticky powder [9, 10]. In extreme environments, such as water shortage, malnutrition, high temperature, the biological crust has strong survival ability, and can gradually improve soil quality and the surrounding environment, plays an important role in the prevention and control of soil erosion and sand-fixation [13]. For the chemical character, it could change the content of soil pH, plant nutrients required important and effectiveness, increase soil organic matter, total nitrogen, total phosphorus and total potassium [11–14]. For the biological character, it could increase the enzyme activity such as soil urease, invertase, catalase and dehydrogenase [15–19]. It has obvious ecological functions in soil and water conservation, improvement, windbreak and sand fixation, and response to global climate change [20], which is of great significance to the sustainable development of desert landscape.

As an important part of desert ecosystem, the formation and development of biological soil crust is one of the main indicators of ecosystem health. However, the natural formation of biological soil crusts is very slow, often taking years or even decades [21]. Therefore, it is imperative for desertification control to accelerate ecological restoration and reconstruction in sandy areas by artificial cultivation and propagation techniques.

The naturally developed BSC has a good sand fixation effect, so can we use artificial cultivation of BSC as a new method to prevent and control desertification? Artificial crust sand-fixation technology is to use BSC fixed sand table and the role of resistance to wind erosion, combined with the traditional biological sand technology, will be the main organisms in the BSC (algae, mosses, and lichens) for artificial cultivation and inoculation to the sand surface, through the maintenance of survival, the surface of the formation of BSC, have the effect of windbreak and sand-fixation, improve the effect of the windbreak and sand-fixation.

A large number of studies and practices have proved that BSC can be cultivated artificially. In addition, the cultured BSCs are characterized by rapid formation. The progress of natural BSC formation is slow in semi-arid and desert areas, and stable BSC can be formed in about 10 years, while artificially cultivated BSC can complete the natural process within one year [22]. In the Tengger Desert, researchers three cyanobacteria (Nostoc sp., Phormidium sp. Scytonemaarcangeli Bornet ex Flahault) were isolated and cultured from native BSCs and then inoculated in quicksand in combination with sand fixers and the superabsorbent polymers. After 1 year, soil hardness increased obviously. The carbohydrate content of newborn BSC, the biomass of cyanobacteria, microbial biomass, soil respiration, carbon fixation and effective quantum yield can be obtained 50% ~ 100% of natural BSC after 20 years of development [23].

Since the 1980s, soil algae biotechnology and microalgae metabolism physiology have been developed rapidly in the world, and a large number of breeding and preservation of fine algae species, and many new technologies have been developed. A large-scale artificial desert algal coating construction (3000 m^2) were carried out in Shapotou area on the southeastern edge of the Tengger Desert in Ningxia Province, in which a large number of Phormidum lucidum were cultivated and directly inoculated on loose sand, and automatic sprinkling irrigation facilities of micro-irrigation were adopted for dune irrigation. The results showed that precipitation and low light intensity could significantly increase the biomass of artificial algae crusting [24]. The feasibility of inoculating cyanobacteria to accelerate soil biological recovery was verified, the results revealed that the inoculation of cyanobacteria increased soil organic carbon and total nitrogen, soil total salt, calcium carbonate and electrical conductivity [25]. The minimum light intensity suitable for the growth of microalgae was determined through indoor culture [26]. Microalgae biomass, microbial biomass and most enzyme activities increased with the development of biological soil crust were found in the Gurbantunggut Desert, Xinjiang [9]. Artificial inoculation is one of the important measures to promote the formation of algae crust in mobile sandy land. Artificial inoculation can make the crust form in quicksand in a short time, and the indoor culture cycle is generally 40–60 days [27].

Existing studies mainly focus on the cultivation process of single microbial algae, and the influence of algal crusts on soil physical and chemical properties and hydrological characteristics [9, 17–19]. However, there is still a lack of understanding on the influence of different ratios of algal crusts on soil physical and chemical and biological properties. In order to provide reference for ecological construction and desertification control in desert areas, different proportions of algal crusting inoculation were carried out in Ulan Buh Desert sample land. In this chapter, the artificial cultivation method of BSC and its effect on soil improvement in sandy areas will be described in detail.

2. Material and methods

2.1 Study site

The research site is located in Dengkou County, Bayannur City, Inner Mongolia, with an altitude of about 1050 m. This region is located in the eastern edge of Ulan Buh Desert, which belongs to temperate continental monsoon climate. The average annual temperature is 7.6°C, and the accumulated temperature during the growing period is about 3100°C. The annual average rainfall is 144.5 mm, and the precipitation is unevenly distributed throughout the year, mainly from June to September. The annual average evaporation is 2397.6 mm, and the frost-free period is 136 days. The main soil type is eolian sandy soil, whose mechanical composition is dominated by fine sand (0.05–0.25 mm), accounting for more than 84%, with little physical clay and coarse sand. The experimental site is located at 106°50'E, 40° 30'N.

2.2 Field survey, lab examination and data analysis

The indoor algae culture experiment began on November 20, 2019. The field algae culture experiment began on January 2, 2020 in Southeastern margin of Ulan Buh Desert, but due to the sand burial, the field experiment did not succeed. After the algae spraying, water was sprayed once a day to keep the surface layer moist. On January 2, 2020, the 0–5 cm soil was sampled and divided into three parts. One part of the fresh soil was sealed in a self-sealing bag and sent to the laboratory at a low temperature in an ice box. The fresh soil was stored in a refrigerator at −80°C for the determination of soil microbial carbon and nitrogen. Part of the fresh soil was sealed in a self-sealing bag and brought back to the laboratory at a low temperature in an ice box. It was stored in a 4°C refrigerator for the determination of ammonium and nitrate nitrogen content. The other parts of the soil were used to determine the organic matter, total carbon, total nitrogen and total phosphorus of the soil after natural air drying, in order to quantify the effects of different ratios of algae crusting on the basic physical and chemical properties of the soil.

Using concentrated sulfuric acid - distillation titration method to measure the total soil nitrogen, using elemental Analyzer to measure the total soil carbon. Soil total phosphorus was determined by alkali fusion - molybdenum antimony anti-spectrophotometry, and microbial carbon and nitrogen were determined by chloroform fumigation extraction method.

One-way ANOVA was used to study the change degree of soil chemical and biochemical indicators of soil samples sprayed with different ratios of algae under 95% confidence interval. If there were significant differences, the least significant difference method (LSD) was used for multiple comparisons. $P < 0.05$ indicated significant differences. Data were collected using SPSS (version 17.0, SPSS Inc., Chicago, IL, USA) and charted using Origin (version 2019, OriginLab Inc., Northampton, MA, USA).

2.3 Artificial culture of algae crust

Algae are the pioneer species for the formation of BSC. As pioneer colonizers, cyanobacteria can grow and reproduce under harsh environmental conditions, such as drought, ultraviolet radiation, poor nutrition, etc. Cyanobacteria can form a fixed quicksand algae crust on the sand surface.

Sand-fixing algae crusts can be formed in a short time by artificial culture. With the development and succession of algeal-sand crusts, and the material input brought by the deposition of fine particles and the accumulation of atmospheric dust, the nutrient enrichment in the sand surface layer was promoted, which created conditions for the reproduction of micro-soil organisms and the colonization of herbaceous plants, and then promoted the desert ecosystem into a virtuous cycle [28].

Algae crust artificial cultivation, is using the principle of algae ecology, physiology, and the theory of biological crust, separating, choosing a natural development formation of the excellent algal crust, after large-scale artificial cultivation, incubated on the surface of the sand to make the rapid formation of crust with algae, bacteria, fungi, algae. The technology mainly includes five aspects [1]:

(1) Isolation, purification and breeding of fine algae species in algae crusts; (2) the scale culture of alga species; (3) Factory/scale production; (4) Incubation in the field; (5) Management and maintenance. The main process and steps of artificial culture of algae crust are as follows (**Figure 1**).

① The algae crusts developed well under natural conditions were adopted. During sampling, the sampling frame is first pressed into the soil forcefully. When the sampling frame is completely put into the soil, the algae crust of about 50 cm^2 in the stainless steel frame is gently taken out with a shovel disinfected with 75% alcohol and put into an envelope for use.

② Sample sieving. Disinfect the gloves with 75% alcohol before operation, and wear gloves; The collected algae crust samples were crushed, passed through a 0.2 mm sieve, and set aside.

③ Cleaning. 20 g of the screened algae crust sample was soaked in 50 mL distilled water and stood for 20 min. The soaked algae crust sample was placed on an ordinary medical gauze and rinsed slowly in distilled water until all the impurities and soil in the sample were removed. The remaining liquid was a mixed algal suspension.

DOI: http://dx.doi.org/10.5772/intechopen.98716

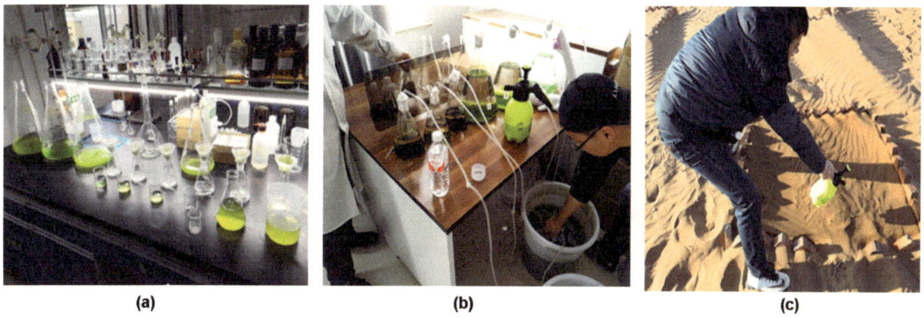

Figure 1.
Flow diagram of artificial algae crust cultivation. (a) Algae cultivation (b) algae propagation (c) field inoculation.

④ The separation of algae species. The algal suspension in step ③ is examined under a microscope. If it is found that a large number of algae need to be separated, it can be separated immediately. If the quantity is small, it should be pre-cultured first, and then separated after increasing.

⑤ Species of algae culture. 2 ~ 5 mL of isolated single or mixed seaweed suspension was taken and added with 200 mL BGI1. The nutrient solution (BG11 culture solution composition and dosage are shown in **Table 1** was placed in a triangular flask. Place the triangular flask on the shaker (Rpm is 140rmin-1) for initial culture, indoor temperature was controlled at 25 ~ 30°C, light intensity was controlled at 600 lx. Indoor culture after 7–10 days, the algae suspension was transferred to a 2-L volumetric flask for preliminary propagation. After propagation for 7 days, the culture medium was transferred into a 50 L plastic bucket, each bucket was equipped with 50 L BG11 culture medium and a set of oxygenation facilities (such as oxygenation pump for domestic ornamental fish); The dosage requirement of algal factory production can be reached after about 7 days of cultivation outdoors.

⑥ The large-scale production of algae. Add the species algae cultured in step ⑤ into the production pool (the fresh weight of species algae added in each pool is about 200 g, and the dry weight is about 2 g). The culture medium used in the incubator was BG11 culture medium, and the water depth was about 0.5 m. In the process of cyanobacteria culture, the growth of cyanobacteria is the best when the water temperature is 25–30°C, and cooling measures should be taken when the water temperature is >40°C (because cyanobacteria will stop growing when the water temperature exceeds

Component	Dosage(mgL^{-1})	Component
$NaNO_3$	1500	the composition of As
$MgSO_4{\bullet}7H_2O$	75	H_3BO_3
$K_2HPO_4{\bullet}3H_2O$	40	$MnCl_2{\bullet}4H_2O$
EDTA-Na_2	1	$Na_2MoO_4{\bullet}2H_2O$
$CaCl_2{\bullet}2H_2O$	36	$ZnSO_4{\bullet}7H_2O$
$C_6H_8O_7$	6	$CuSO_4{\bullet}5H_2O$
$C_6H_8FeNO_7$	6	
As	1 ml/L	
Na_2CO_3	20	

Table 1.
Composition and dosage of BG11 medium.

40°C). The light intensity is controlled at 15000 lx. Harvest will be carried out after 7 to 12 days of culture of cyanobacteria (about 7 days in July and August in summer, 10–12 days in April and June in spring and 10–12 days in September and October in autumn). Determination of harvest period: it was determined according to the growth curve of cyanobacteria. When the growth rate or increase amount of cyanobacteria began to decline (the growth amount began to decline after reaching the maximum), the harvest was carried out. Before harvest, the liquid in the culture tank was left standing for 6 ~ 8 h. The culture solution is then discharged into an empty pool. (After one culture, the algae can still grow, but the yield is reduced.)30% ~ 40%. During the culture process, green algae from the air will enter the culture pond, leading to increased competition between algae and cyanobacteria. When the water depth in the pool is about 0.5 cm, the algal fluid is collected into a plastic bucket for use.

2.4 Field application of algae crust

Field inoculation: the collected algal solution was evenly injected into the sandy land surface with a sprayer, and the amount of algal solution was 3 g dry weight m^{-2}. For the first 10 days after inoculation, spray water every 2 days (the surface is moist), stop watering during rain. A month later, to measure the formation of algae crust, the topsoil area of 5 cm^2 and the thickness of about 5 mm were used for detection (due to the main algae distributed within 5 mm of the soil surface, so each sampling should be no more than 5 mm), and the content of chlorophyll a and chlorophyll b were measured.

3. Effect of algal crusts on soil improvement

3.1 Effects of different treatments on soil physical indexes

The soil bulk density after spraying algae culture solution in different proportions was shown in **Figure 2**. After spraying different proportions of algae culture solution, the bulk density of the 0–5 cm surface soil tended to decrease as a whole, among which, the soil bulk density of spraying mixed algae Microcolus vaginatus:

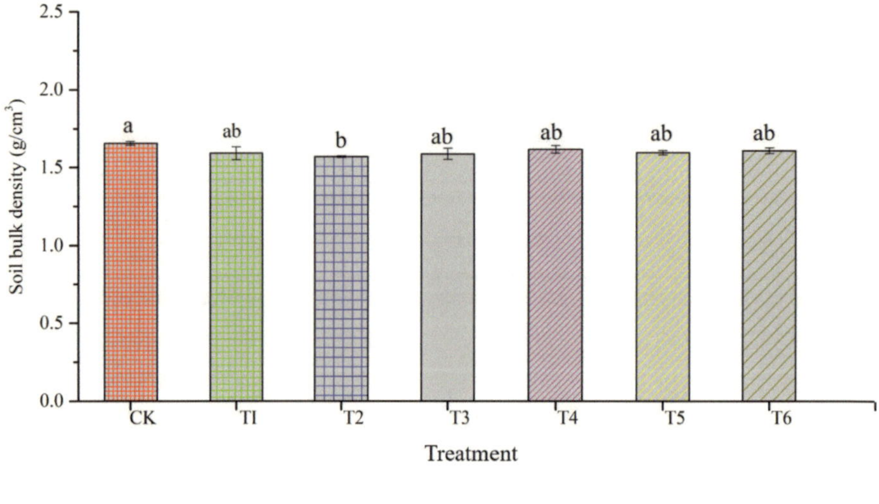

Figure 2.
Effects of different treatments on soil physical indexes. CK T1: Microcolus vaginatus: Scytonema sinense: Phormidum lucidum 1:1:1; T2: Microcolus vaginatus: Scytonema sinense: Phormidum lucidum 3:2:1; T3: Microcolus vaginatus: Scytonema sinense 5:1; T4: Microcolus vaginatus; T5: Scytonema sinense; T6: Phormidum lucidum.

Scytonema sinense: Phormidium lucidum 3:2:1 significantly decreased ($P < 0.05$). In conclusion, spraying different proportion of algal culture solution can reduce the soil bulk density, and the mixed algae with the ratio of Microcolus vaginatus: Scytonema sinense: Phormidium lucidum 3:2:1 has the best culture status. In the study of biological soil crusts in the Horqin Sandy Land of Inner Mongolia, algae crusts can significantly reduce the surface soil bulk density [21].

3.2 Effects of different treatments on soil chemical indexes

Accumulation of soil carbon and nitrogen is regarded as an indicator of soil fertility and productivity [29]. Algae spraying in different proportions had different effects on soil organic carbon. Among them, the ratio of single algae species Scytonema sinense and Phormidium lucidum, and mixed algae species Microcolus vaginatus: Scytonema sinense 5:1 did not change soil organic carbon content, while the ratio of single algae species Microcolus vaginatus, Microcolus vaginatus: Scytonema sinense: Phormidium lucidum 1:1:1 showed a trend of increasing soil organic carbon content. The soil total carbon content was significantly increased by 72.6% ($P < 0.05$) by spraying mixed algae with the ratio of Microcolus vaginatus: Scytonema sinense: Phormidium lucidum 3:2:1 (**Figure 3**).

Previous studies have shown that biological soil crusts can fix atmospheric carbon and nitrogen [30, 31]. Fixed carbon and nitrogen are released into the surrounding environment and used by other organisms such as vascular plants, fungi and bacteria [32]. The study of the algal crust under the canopy of Artemisia sphaerica vegetation and the organic matter in the underlying soil in Mu Us Sandy Land found that most of the algal crust developed stably between canopies, and most of the algal crust under the canopy was at the early or middle stage of development, and the content of organic matter in the intercanopies and underlying layer increased significantly [33]. Nitrogen fixing cyanobacteria can fix atmospheric nitrogen and increase soil organic matter content. The nitrogen-fixing activity of algal crusts in the artificial vegetation sand-fixing area and the natural vegetation area were compared, the results indicated that the nitrogen-fixing activity of algal crusts in the artificial vegetation area increased significantly with the extension of vegetation restoration time [34]. The results showed that the development of algal crusts promoted soil nitrogen content and soil development after the implementation of artificial vegetation restoration measures.

The difference in total nitrogen content is an indication of different nitrogen inputs [13]. Soil total nitrogen content and soil organic carbon content had a similar response trend. Neither single algal culture medium could significantly increase soil total nitrogen content, but the soil total nitrogen content in the ratio of Microcolus

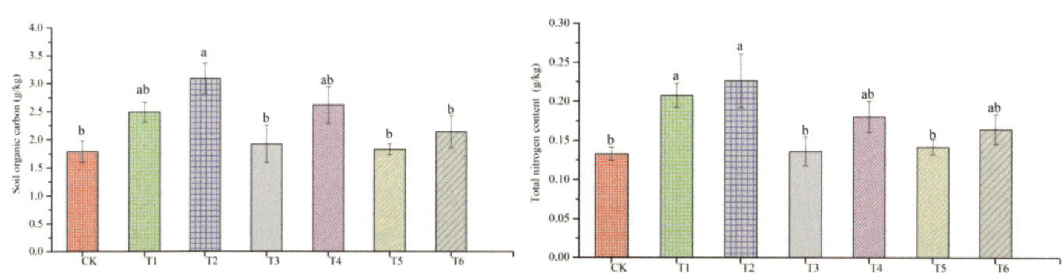

Figure 3.
Effects of different treatments on soil chemical indexes. CK T1: Microcolus vaginatus: Scytonema sinense: Phormidium lucidum 1:1:1; T2: Microcolus vaginatus: Scytonema sinense: Phormidium lucidum 3:2:1; T3: Microcolus vaginatus: Scytonema sinense 5:1; T4: Microcolus vaginatus; T5: Scytonema sinense; T6: Phormidium lucidum.

vaginatus: Scytonema sinense: Phormidum lucidum 3:2:1 was significantly increased ($P < 0.05$), with the increase ratios of 56.2% and 70.3%, respectively.

All the algal culture media had no effect on soil total P content. Carbon input is an important ecological function of biological soil crust in arid regions, and the ability of carbon fixation is affected by the development degree of biological soil crust. In well-developed biological soil crusts, the carbon fixation rate is about two times higher than that in poorly developed soil crusts, which is mainly attributed to the increase of chlorophyll a in well-developed biological soil crusts [35].

Most of the nitrogen fixed by biological soil crusts can be immediately released into the soil [36]. At low soil moisture content, biological soil crusts had no effect on soil organic matter and soil total nitrogen content, while at high soil moisture content, Biological soil crusts can significantly increase soil organic carbon and soil total nitrogen content at 0–5 cm depth [37]. This may be because the available water under the surface of the soil is too high, and the water condition is high enough to activate carbon fixation components [38] and drive photosynthesis to produce ATP and carbohydrates for nitrogen fixation [39].

3.3 Effects of different treatments on soil biological indexes

Soil microbial biomass fluctuated between 0.29 and 2.02 g C m^{-2} in soil biological crust, and its value was significantly correlated with the development degree of soil crust [40]. Increasing organic matter and polysaccharides offers a plentiful carbon source for microorganisms and invertase, resulting in an increase of microbial biomass C [41, 42]. Increase of microalgal biomass is helpful to improve organic C and available P. The release of carbonic acid by the algal cells can accelerate the weathering of minerals, hence make improvement of inorganic ions [43].

All medium significantly increased the content of soil microbial biomass carbon ($P < 0.05$), but there was no significant difference among different medium ($P > 0.05$).

Except for Scytonema sinense, other algal culture media significantly increased the soil single species Scytonema sinense and Phormidum lucidum significantly increased soil microbial biomass phosphorus ($P < 0.05$), and other species had an increasing trend, but the statistical difference was not significant (**Figure 4**).

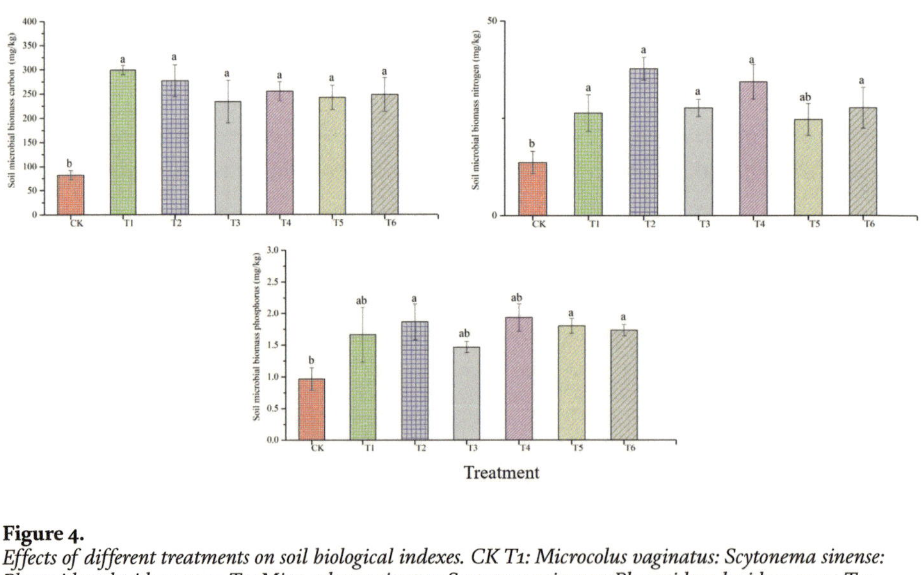

Figure 4.
Effects of different treatments on soil biological indexes. CK T1: Microcolus vaginatus: Scytonema sinense: Phormidum lucidum 1:1:1; T2: Microcolus vaginatus: Scytonema sinense: Phormidum lucidum 3:2:1; T3: Microcolus vaginatus: Scytonema sinense 5:1; T4: Microcolus vaginatus; T5: Scytonema sinense; T6: Phormidum lucidum.

3.4 Existing problems and suggestions

As a new method to prevent and control desertification, artificial cultivation of algae crust has made substantial progress and breakthrough in the practice of preventing and controlling desertification. However, the following two aspects need to be further discussed: sand burial is one of the most common disturbance factors of desert ecosystem in sand area, especially in arid desert ecosystem with frequent eolian sand activities. Sand burial affected the growth and survival of BSC by changing the light, temperature and soil physical and chemical properties of BSC habitat. Especially in the early stage of the construction of artificial algae crust, the survival ability of algae is relatively weak, and the existence of sand crust seriously threatens the further development and formation of algae crust. Artificial crust is more fragile, and its resistance to adversity is weak. Therefore, after the construction of artificial algae crust is completed, how to prevent and control the damage of sand burial to it. It is a problem to be faced in constructing artificial algae crust. The key links of technology still need further optimization and innovation, still need to be improved and perfected through a lot of scientific research and production practice. In addition, the early artificial hydration measures after inoculation are conducive to the normal synthesis and metabolism of extracellular polysaccharide, and the accumulation of extracellular polysaccharide is helpful to restore the drought-tolerant ability of algae and adapt to the external drought environment, thus increasing the biomass of algae, improving the drought-tolerant ability of algae and promoting the formation of artificial algae crust. Therefore, early water acquisition is the key factor for the successful formation of artificial algae crust. However, water is the most important ecological limiting factor in arid areas. If water is replenished continuously in the early stage, it cannot be realized in large-scale desertification prevention and control practices. Therefore, how to cultivate the algae with stronger drought resistance and more suitable for the formation of artificial algae crust is the key technology to be broken through in the construction of artificial algae crust.

4. Conclusions

This study shows that different proportions of algae crusts can improve the physical, chemical and biological properties of desert soil to different degrees. The results showed that the soil bulk density in the 0–5 cm surface layer had a decreasing trend after spraying different proportion of algae culture solution, and the soil bulk density in the mixed algae Microcolus vaginatus: Scytonema sinense: Phormidum lucidum 3:2:1 was significantly decreased ($P < 0.05$). The soil organic carbon content was increased with the ratio of single species Microcolus vaginatus, Microcolus vaginatus: Scytonema sinense: Phormidum lucidum: 1:1:1. The soil total carbon content was significantly increased by 72.6% ($P < 0.05$) when the ratio of mixed species Microcolus vaginatus: Scytonema sinense: Phormidum lucidum: 3:2:1. The soil total nitrogen content was significantly increased with the ratio of Microcolus vaginatus: Scytonema sinense: Phormidum lucidum 3:2:1 ($P < 0.05$), increasing by 56.2% and 70.3%, respectively; All medium significantly increased the soil microbial biomass carbon content ($P < 0.05$). And artificial cultivation of algae crusts can greatly shorten the formation time of algae crusts. The results in this study highlight the significant role of BSCs for soil improvement in semiarid and arid areas. Therefore, when carrying out ecological construction in arid desert, it is necessary to fully consider the difference of effects of different ratio of algal solution, so as to achieve the optimal soil nutrient improvement effect of biological crusts.

Acknowledgements

We acknowledge financial support provided by the National Key Research and Development Program of China (Project No. 2018YFC0507101-04) and the Strategic Priority Research Program of the Chinese Academy of Sciences, Grant No.XDA 26020103. Thanks to Xiaoliang An and Yingyu Hu for their help in the experiment and data processing.

Conflict of interest

The authors declare no conflict of interest.

Author details

Jinchao Feng[1*], Wei Li[1] and Linna Ma[2]

1 Institute of Desertification Studies, Chinese Academy of Forestry, Beijing, China

2 State Key Laboratory of Vegetation and Environmental Change, Institute of Botany, Chinese Academy of Sciences, Beijing, China

*Address all correspondence to: fengjinchao@caf.ac.cn

References

[1] Li XR, Hui R, Zhao Y. Eco-physiology of Biological Soil Crusts in Desert Regions of China. Beijing, Higher Education Press, 2016, 386-388

[2] Wei JC. Desert biological carpet engineering - a new way to control arid desert. Arid Zone Research, 2005, 22(3), 287-288. doi: CNKI:SUN:GHQJ.0.2005-03-001

[3] Li XR, Tan HJ, Hui R, et al. Study on biological soil crust in desert and sandy land of China [J]. Chinese Science Bulletin, 2018,63(23). doi: 10.1360/N972018-00390

[4] West N E. Structure and Function of Microphytic Soil Crusts in Wildland Ecosystems of Arid to Semi-arid Regions[J]. Advances in Ecological Research, 1990,20:179-223.doi: 10.1016/S0065-2504(08)60055-0

[5] Eldridge D, Greene R. Microbiotic soil crusts - a review of their roles in soil and ecological processes in the rangelands of Australia[J]. Soil Research, 1994,32(3): 389-415.doi: 10.1071/sr9940389

[6] Meng J, Bu CF, Zhang XC, et al. Developmental characteristics of biological soil crust under different vegetation types in wind-water erosion crisscross region, Northern Shaanxi Province, China[J], Journal of Northwest Forestry University,2011,26(4):41-46.doi: CNKI: SUN: XBLX. 0.2011-04-010

[7] Lv YZ, Yang PG. The effects of desert crust on the character of soil water. Journal of Arid Land resources and environment [J]. Journal of Arid Land Resources and Environment, 2004, 18(002):76-79. doi: 10.3969/j.issn.1003-7578.2004.02.015

[8] Belnap J. The potential roles of biological soil crusts in dryland hydrologic cycles[J]. Hydrological Processes, 2010,20(15):3159-3178.doi: 10.1002/hyp.6325

[9] Zhang B C, Zhou X B, Zhang Y M. Responses of microbial activities and soil physical-chemical properties to the successional process of biological soil crusts in the Gurbantunggut Desert, Xinjiang[J]. Journal of Arid Land, 2015,7(001):101-109.doi: 10.1007/s40333-014-0035-3

[10] Gao L, Bowker M A, Xu M et al. Biological soil crusts decrease erodibility by modifying inherent soil properties on the Loess Plateau, China[J]. Soil Biology and Biochemistry, 2017,105:49-58. doi: 10.1016/j.soilbio.2016.11.009Get rights and content

[11] Zhao YG, Xu MX, Wang QJ, et al. physical and chemical properties of soil bio-crust on rehabilitated grassland in hilly loess plateau of china. [Z]. 2006: 17, 1429-1434.doi: CNKI: SUN: YYSB.0.2006-08-013

[12] Zhao HL, Guo ZR, Zhou RL, et al. Effects of vegetation cover on physical and chemical properties of bio-crust and under-layer soil in horqin sand land.[Z]. 2009: 20, 1657-1663.doi: http://ir.casnw.net/handle/362004/11401

[13] Gao LQ, Bowker MA, Xu MX, et al. Biological soil crusts decrease erodibility by modifying inherent soil properties on the Loess Plateau, China. Soil Biol Biochem, 2017, 105: 49-58.doi: 10.1016/j.soilbio.2016.11.009

[14] Veste M. The importance of biological soil crusts for rehabilitation of degraded arid and semi-arid ecosystems[J]. Science of Soil and Water Conservation, 2005,3:42-47.

[15] Zhang BC, Zhou XB, Zhang YM. Responses of microbial activities and soil physical-chemical properties to the successional process of biological soil crusts in the Gurbantunggut Desert, Xinjiang. J Arid Land, 2014, 7: 101-109. doi: 10.1007/s40333-014-0035-3

[16] Niu J, Yang K, Tang Z, et al. Relationships between soil crust development and soil properties in the desert region of North China. Sustainability, 2017, 9: 725. doi: 10.3390/su9050725

[17] Zhang ZS, Dong XJ, Liu YB, et al. Soil oxidases recovered faster than hydrolases in a 50-year chronosequence of desert revegetation.Plant Soil, 2012, 358: 275-287. doi: 10.1007/s11104-012-1162-2

[18] Liu YM, Yang H Y, Li X R, et al. Effects of biological soil crusts on soil enzyme activities in revegetated areas of the Tengger Desert, China. Appl Soil Ecol, 2014, 80: 6-14. doi: 10.1016/j.apsoil.2014.03.015

[19] Zhou XB, Zhang YM. Temporal dynamics of soil oxidative enzyme activity across a simulated gradient of nitrogen deposition in the Gurbantunggut Desert, Northwestern China. Geoderma, 2014, 213: 261-267. doi: 10.1016/j.geoderma.2013.08.030

[20] Hu CX, Liu YD, Song LR. new development of soil algae research [J]. Acta Hydrobiologica Sinica, 2002(05): 521-528.doi: 10.3321/j.issn:1000-3207.2002.05.018

[21] Thiet R K, Boerner R E J, Nagy M et al. The Effect of Biological Soil Crusts on Throughput of Rainwater and N into Lake Michigan Sand Dune Soils[J]. Plant and Soil, 2005,278(1-2):235-251.

[22] Chen LZ, Liu YD, Song LC. The function of exopolysaccharides of microcoleus in the formation of desert soil. Acta Hydrobiologica Sinica, 2002,26(2):1 55-159. doi: 10.3321/j.issn:1000-3207.2002.02.008

[23] Park CH, Li XR, Zhao Y, et al. Rapid development of cyanobacterial crust in the field for combating desertification. PLoS One, 2017,12: e0179903. doi: 10.1371/journal.pone.0179903

[24] Chen L, Xie Z, Hu C et al. Man-made desert algal crusts as affected by environmental factors in Inner Mongolia, China[J]. Journal of Arid Environments, 2006,67:521-527.doi: 10.1016/j.jaridenv. 2006.02.018

[25] Wang W, Liu Y, Li D et al. Feasibility of cyanobacterial inoculation for biological soil crusts formation in desert area[J]. Soil Biology & Biochemistry, 2009,41(5):926-929.doi: 10.1016/j.soilbio.2008.07.001

[26] Lan S, Wu L, Zhang D et al. Effects of light and temperature on open cultivation of desert cyanobacterium Microcoleus vaginatus[J]. Bioresource Technology, 2015,182:144-150.doi: 10.1016/j.biortech.2015.02.002

[27] Wang XR, Zhao YG, Wang Y. A review on studies of moss crust artificial cultivation in arid and semi-arid region. Journal of Northwest Forestry University [J]. 2014,29(006):66-71.doi: 10.3969/j.issn.1001-7461.2014.06.13

[28] Hu CX, Liu YD, Song LR, et al. Species composition and distribution of algae in semi-desert algal crusts. Chinese Journal of Applied Ecology, 2000,11(1):61-65. doi: CNKI:SUN:YYSB.0.2000-01-015

[29] Guo Y, Zhao H, Zuo X et al. Biological soil crust development and its topsoil properties in the process of dune stabilization, Inner Mongolia, China[J]. Environmental Geology, 2008.doi: 10.1007/s00254-007-1130-y

[30] Johnson S L, Neuer S, Garcia-Pichel F. Export of nitrogenous compounds due to incomplete cycling within biological soil crusts of arid lands[J]. Environmental Microbiology, 2010,9(3):680-689.doi: 10.1111/j.1462-2920.2006.01187.x

[31] Veluci, RM, Neher et al. Nitrogen fixation and leaching of biological soil crust communities in mesic temperate soils[J]. MICROBIAL ECOL, 2006.doi: 10.1007/s00248-005-0121-3

[32] Belnap J, Lange O L. Biological Soil Crusts: Structure, Function, and Management[J]. 2003, doi:10.1007/978-3-642-56475-8.

[33] Dong JW, Li YP, Li XK, et al. Effects of Vegetation on Biocrusts and the Underlying Soil Nutrients in Mu Us Sandland[J]. Research of Soil and Water Conservation, 2019,026(002):112-117. doi: CNKI:SUN:STBY.0.2019-02-019

[34] Su YG, Li XR, Jia XH, et al. Nitrogenase Activity Chronosequence of Algal Crusts in Temperate Desert Region[J]. Journal of Desert Research. 2012,32(002):421-427. doi: http://ir.casnw.net/handle/362004/11567

[35] Housman D C, Powers H H, Collins A D et al. Carbon and nitrogen fixation differ between successional stages of biological soil crusts in the Colorado Plateau and Chihuahuan Desert[J]. Journal of Arid Environments, 2006,66(4):620-634. doi: 10.1016/j.jaridenv.2005.11.014

[36] Pendleton R L, Pendleton B K, Howard G L et al. Growth and Nutrient Content of Herbaceous Seedlings Associated with Biological Soil Crusts[J]. Arid Soil Research & Rehabilitation, 2003,17(3):271-281. doi: 10.1080/15324980301598

[37] Shuqin G, Xuehua Y, Yu C et al. Effects of biological soil crusts on profile distribution of soil water, organic carbon and total nitrogen in Mu Us Sandland, China[J]. Journal of Plant Ecology, 2010(4):279-284. doi: 10.1093/jpe/rtq015

[38] Thomas A D, Hoon S R, Linton P E. Carbon dioxide fluxes from cyanobacteria crusted soils in the Kalahari[J]. Applied Soil Ecology, 2008,39(3):254-263. doi:10.1016/j.apsoil.2007.12.015

[39] Bergman B. N2 Fixation by non-heterocystous cyanobacteria[J]. Fems Microbiology Reviews, 1997,19(3): 139-185. doi: 10.1111/j.1574-6976.1997.tb00296.x

[40] Yu J, Kidron G J, Pen-Mouratov S et al. Do development stages of biological soil crusts determine activity and functional diversity in a sand-dune ecosystem?[J]. Soil Biology & Biochemistry, 2012,51(none):66-72. doi:10.1016/j.soilbio.2012. 04. 007

[41] Katsalirou E, Deng S, Nofziger DL, et al. Long-term management effects on organic C and N pools and activities of C-transforming enzymes in prairie soils[J]. European Journal of Soil Biology, 2010, 46(5):335-341. doi: 10.1016/j.ejsobi.2010. 06.004

[42] Zhou XB, Zhang YM, Downing A. Non-linear response of microbial activity across a gradient of nitrogen addition to a soil from the Gurbantunggut Desert, northwestern China[J]. Soil Biology & Biochemistry, 2012, 47(none):67-77. doi: 10.1016/j.soilbio. 2011. 05. 012

[43] Killham K. 1994. Soil Ecology. Cambrige: Cambridge University Press,40-61.

Chapter 5

Water Source of Six Woody Plants in Different Habitats on Desertified Land of Ordos Plateau, Semi-Arid China

Yajuan Zhu

Abstract

Water and soil erosion and sandy desertification are two mainly land desertification types on eastern and southern Ordos Plateau, north China. *Hippophae rhamnoides*, *Armeniaca sibirica* and *Pinus tabuliformis* are three woody plants for soil and water conservation on loess slope. *Sabina vulgaris*, *Artemisia ordosica* and *Salix psammophila* are three shrubs for sand control on sand dune. Water source of six woody plants were investigated by stable isotope technology. The results showed that the $\delta^{18}O$ of shallow soil water was similar to that of rainwater in July and September in two habitats. Both of six woody plants in two habitats mainly used shallow soil water in May. However, three shrubs on sand dune mainly used both of shallow and deep soil water in July and September. Three woody plants on loess slope mainly used rainwater or deep soil water in July and September. Therefore, six woody plants utilized different depths of soil water or rain water based on their availability in different seasons, which is an adaptive strategy to the semiarid climate on Ordos Plateau.

Keywords: land desertification control, rain water, soil water, stable oxygen isotope, water source

1. Introduction

In semi-arid region, water is the key factor for the survival and succession of plant community [1]. The use of limited water resource concerns not only plant survival, but also interspecific interaction and community dynamics. Since there is generally no stable isotope fractionation during water uptake by root system and water transportation before arriving leaf, water source of a plant could be identified by comparing stable isotope of xylem water and potential water sources [2]. The potential water sources for plant species are shallow soil water recharged by rainwater, deep soil water recharged by rainwater, snow water or groundwater in semi-arid region [3–10]. Different life form plants usually used different water source, which related to their root types. Generally, tree and shrub with deep root system could use deep soil water or groundwater [6]. Woody species with dimorphic root system may use different depths of soil water or groundwater simultaneously [9]. And shallow rooted shrub and perennial grass only used shallow or middle soil water [11]. Water source of plant species in semi-arid region was affected by many environmental factors, like season [6, 7], annual variance of climate [8] and habitat heterogeneity [12].

Ordos Plateau is located in the middle reaches of Yellow River, which is the ecotone between Loess Plateau and Mongolian Plateau, with the total area of 1.30×10^5 km^2. The eastern part is hilly gully area with loess hill and valley, the southern part is Mu Us Sandy Land with fixed and semi-fixed sand dune, and the northern part is Hobq Desert with moving sand dune [13]. The elevation increases from 774 m in southeast to 2148 m in northwest [14]. From southeast to northwest, the annual precipitation decreased from 400 mm to 200 mm; meanwhile, the natural vegetation varies from forest grassland, grassland, and sandy land to desert [13]. In the last decades, land desertification was severely on Ordos Plateau. From 1994 to 2000, land desertification area increased for 1.90×10^4 km^2, focused on west Hobq Desert and south Mu Us Sandy Land [15]. In these years, many measures was taken to increase vegetation coverage, decrease sand storm hazard and improve ecological environment, such as air seeding, fence and afforestation.

In eastern Ordos Plateau, the hilly gully area is one of the most severely soil and water erosion region in the middle reaches of Yellow River, which is sand stone covered by loess and contributed up to 25% of the total course sediment for the lower reaches [16]. Both of engineering measure and vegetation restoration was carried out to control soil and water erosion, such as plant trees and shrubs, including *Hippophae rhamnoides*, *Pinus tabuliformis* and *Armeniaca sibirica*.

Mu Us Sandy Land is one of the four sandy land in China, which located in central Inner Mongolia, Northern Shaanxi and Northeast Ningxia. The total area is 4.22×10^4 km^2, with the elevation varies from 1000 to 1600 m [17]. The main natural vegetation are forest steppe, steppe, and shrub sandy land and desert steppe. Vegetation growth was improved in Mu Us Sandy Land, which were resulted by climate change and human activity [18]. The dominant species of sand control are *Sabina vulgaris*, *Artemisia ordosica*, *Salix psammophila* and *Caragana intermedia* in shrub sandy land [19].

Previous studies in Mu Us Sandy Land showed that *S. vulgaris* mainly used soil water within 1.5 m and groundwater, whereas *A. ordosica* mainly used shallow soil water within 50 cm [5]. Moreover, *A. ordosica* mainly used deep soil water recharged by 65 mm rainstorm in summer, *Cynanchum komarovii* mainly used middle rain of 10–20 mm, whereas *Stipa bungeana* mainly used shallow soil water recharged by small rain [3]. However, we still do not know the seasonal dynamic of water source for these dominant shrubs in Mu Us Sandy Land. Moreover, there is few report about water source of woody species in hilly gully area of Ordos Plateau. Therefore, seasonal dynamic of water source of six woody plants was explored by stable isotope technology in two different habitats of Ordos Plateau. The purpose of this study was to understand how these trees and shrubs adapt to the semi-arid climate by adjusting water source in the growing season. The results would give theoretical supports for ecological forestry engineering, including Natural Forest Protection, Grain for Green and Three Norths Shelterbelt Program.

2. Water source of three woody species on loess slope

2.1 Three soil and water conservation woody species on loess slope

This study was conducted in Soil and Water Conservation Park of Jungar Banner, Ordos City, Inner Mongolia. The banner has a temperate continental climate, which mean air temperature varies from 6.2°C to 8.7°C, mean annual precipitation is 400 mm, potential evapotranspiration is 2093 mm and forest free days are 145 d [20]. The natural vegetation is steppe dominated by *Stipa bungeana*, whereas few *P. tabuliformis*, *Juniperus rigida* and *Platycladus orientalis* are distributed on hill slope.

Hippophae rhamnoides (sea buckthorn) is a small tree or shrub with the height of 1–5 m. It often inhabits in hill ridge, valley, dry river bed or slope with rock, sandy loam or loam, which distributes in Hebei, Inner Mongolia, Shanxi, Shaanxi, Gansu, Qinghai and West Sichuan. This shrub was widely used on Loess Plateau as a soil and water conservation species [21]. It grows fast, resistant to drought and could fix nitrogen.

Pinus tabuliformis (Chinese pine) is an evergreen tree with the height of 25 m and the diameter at breast height of 1 m. It is the dominant species in coniferous forest with elevation varies from 100 to 2600 m, which distributes in south Jilin, Liaoning, Hebei, Henan, Shandong, Shanxi, Inner Mongolia, Shaanxi, Ningxia, Qinghai and Sichuan. It is used as a sand binding and soil and water conservation trees in north China [22].

Armeniaca sibirica (wild apricot) is a small tree or shrub with the height of 2–5 m. It mainly inhabits on dry slope, hill steppe or mixed with deciduous forest, which distributes in Heilongjiang, Jilin, Liaoning, Inner Mongolia, Gansu, Hebei and Shanxi, and also in East and Southeast Mongolia, Far East and Siberia of Russia [23]. It is cold-resistant and was used as a soil and water conservation species in Northeast and North China.

In the study site, three woody species was planted in 2013 with density of 1000 plants per hectare. The mean height of *H. rhamnoides*, *P. tabuliformis* and *A. sibirica* were 2.10, 2.14 and 2.71 m, respectively. The main water source of three woody species was measured by comparing their xylem water with different water sources, e.g. rainwater and soil water in 10, 25, 50, 75 and 100 cm. Lignified, two years old twigs of shrubs were collected and the bark was removed with scissor, then the xylem was sampled. Soil and xylem were placed in 8 mL glass vial, sealed with Parafilm, and stored in a medical cool box. Water in soil and xylem was vacuum-extracted and their $\delta^{18}O$ value was measured with a Flash 2000HT elemental analyzer and a Finnigan MAT 253 mass spectrometer. Meteorological data was obtained from Meteorology Bureau of Dongsheng District, Ordos City, which is 50 km west to the study site. The contribution of different water sources to their total water use was analyzed by Iso-source 1.3.1 software [24]. The input data of the model were $\delta^{18}O$ value of xylem water and potential water source, e.g. soil water in different depths or rain water. The results of water use ratio to different sources were expressed as mean ± SD.

2.2 Precipitation of study site in the growing season of 2018

The total precipitation of Dongsheng District was 385.00 mm in the growing season of 2018 (**Figure 1**), which was slightly lower than annual mean precipitation of 400 mm. The maximal daily precipitation was 52.0 mm and occurred on May 19, the next value was 48.8 mm on July 19. The monthly precipitation were 32.0, 77.9, 15.5, 118.3, 85.0 and 54.3 mm from April to September.

2.3 Stable oxygen isotope of xylem water of three woody species, soil water and rainwater on loess slope

On May 12, stable oxygen isotope ratio of *H. rhamnoides* and *P. tabuliformis* xylem water were closer to soil water in 10 cm, whereas stable oxygen isotope ratio of *A. sibirica* xylem water was closer to soil water in 10–25 cm. On July 15, stable oxygen isotope ratio of 10–25 cm soil water was closer to rainwater. Stable oxygen isotope

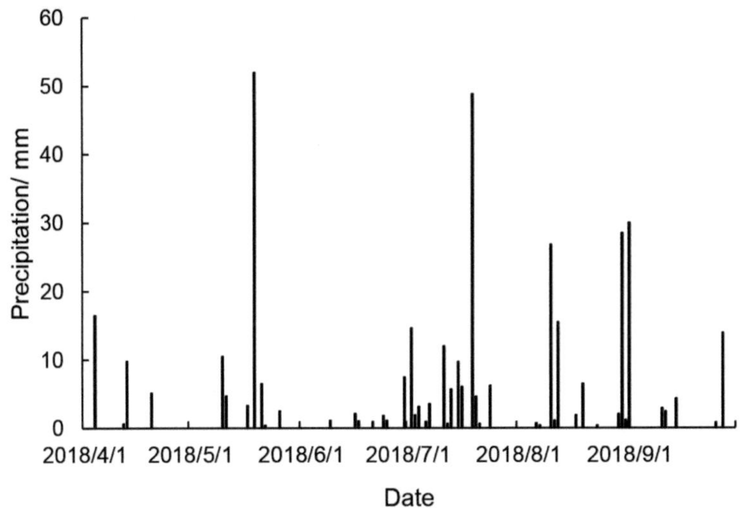

Figure 1.
Daily precipitation of study site from April to September of 2018.

ratio of *H. rhamnoides*, *P. tabuliformis* and *A. sibirica* xylem water were closer to soil water in 10–100 cm, 25–50 cm and 25–100 cm, respectively; moreover, stable oxygen isotope ratio of *P. tabuliformis* and *A. sibirica* xylem water were also closer to rainwater. On September 23, stable oxygen isotope ratio of *H. rhamnoides* xylem water was closer to soil water in 25–100 cm, whereas stable oxygen isotope ratio of xylem water of other two woody species were closer to soil water in 10–100 cm (**Figure 2**).

2.4 Contribution of different depth of soil water and rainwater to the water source of three woody species on loess slope

Iso-source analysis showed that *H. rhmnoides* mainly used 10 cm soil water on May 12, which accounted for 88.5% of its total water source. On July 15, it mainly used 10–25 cm soil water and rainwater, which accounted for 44.6% and 35.4% of its total water source. On September 23, it mainly used 25 cm and 75–100 cm soil water, which accounted for 88.9% of its total water source (**Table 1**).

Iso-source analysis showed that *P. tabuliformis* mainly used 10 cm soil water on May 12, which accounted for 94.0% of its total water source. On July 15, it mainly used rainwater, which accounted for 93.7% of its total water source. On September 23, it mainly used 10 cm and 50–75 cm soil water, which accounted for 84.5% of its total water source (**Table 2**).

Iso-source analysis showed that *A. sibirica* mainly used 10 cm soil water on May 12, which accounted for 91.6% of its total water source. On July 15, it mainly used 25–100 cm soil water and rainwater, which accounted for 55.9% and 36.8% of its total water source. On September 23, it evenly used 10–100 cm soil water (**Table 3**).

Three woody species on loess slope selected different water source in the growing season, which is an adaptive strategy to semi-arid environment. They mainly used shallow soil water recharged by spring rain. However, there are interspecific difference of water source in summer. For *H. rhmnoides*, it mainly used shallow soil water and rainwater. For *P. tabuliformis*, it mainly used rainwater. However, for *A. sibirica*, it mainly used middle and deep soil water and rainwater. In autumn, *H. rhmnoides* mainly used shallow and deep soil water, *P. tabuliformis* mainly used shallow and middle soil water. However, *A. sibirica* evenly used different depths soil water. The interspecific difference in water source of three woody species is related

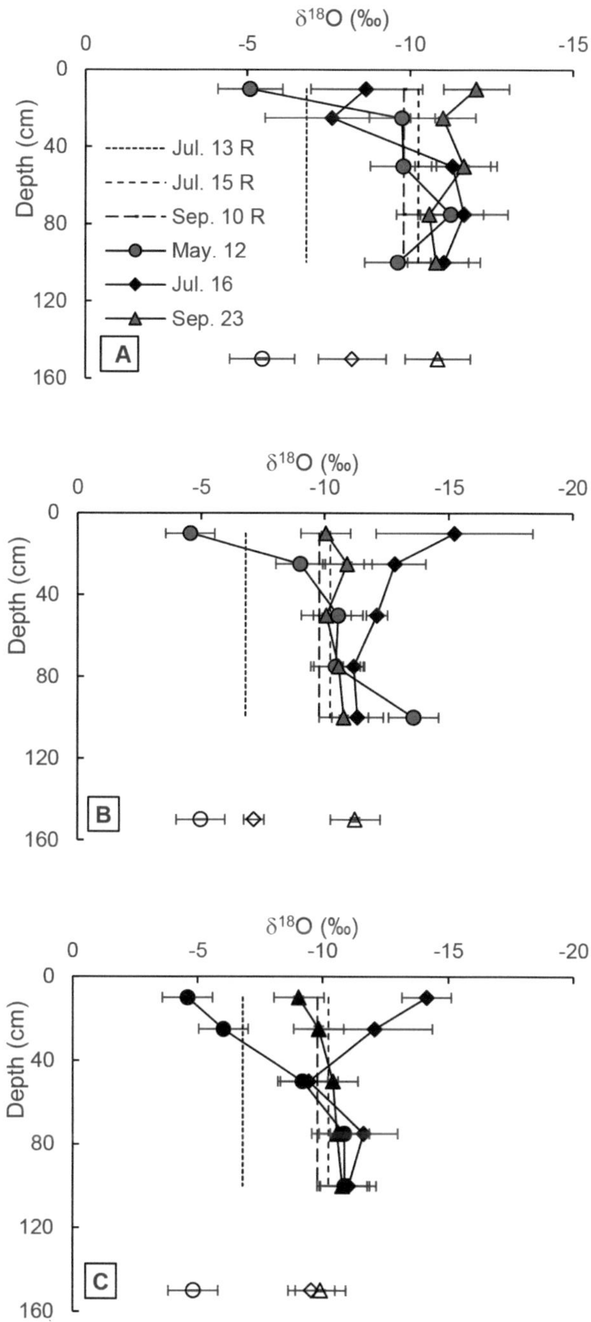

Figure 2.
Stable oxygen isotope ratio of xylem water, soil water and rainwater of Hippophae rhamnoides (A), Pinus tabuliformis (B) and Armeniaca sibirica (C) in hilly gully area. Dark symbol is soil water, white symbol is xylem water of three woody species.

to their life form and root distribution. Other shrubs on Loess Plateau also changed their water source. For example, *Caragana korshinskii* and *Salix psammophila* used 40–80 cm soil water in the drought season. However, they mainly used 0–40 cm soil water in the rain season [25]. *Armeniaca sibirica* in the mixed plantation used more shallow soil water than that in the pure plantation. However, *Robinia pseudo-acacia* always used shallow and middle soil water in different plantation types [26].

Water source	May 12	Jul. 15	Sep. 23
10 cm soil water	88.5 ± 0.7	16.4 ± 13.7	4.6 ± 3.8
25 cm soil water	3.1 ± 2.8	28.2 ± 20.5	17.0 ± 13.3
50 cm soil water	3.0 ± 2.8	6.7 ± 5.6	6.4 ± 5.2
75 cm soil water	1.9 ± 2.2	6.2 ± 5.2	41.4 ± 18.3
100 cm soil water	3.4 ± 2.6	7.1 ± 0.6	30.5 ± 22.8
Rainwater	—	35.4 ± 17.1	—

Table 1.
Contribution of different depths of soil water and rainwater to the water source of Hippophae rhamnoides in hilly gully area (%, mean ± SD).

Water source	May 12	Jul. 15	Sep. 23
10 cm soil water	94.0 ± 0.9	0.6 ± 0.8	36.8 ± 21.0
25 cm soil water	1.8 ± 2.3	1.2 ± 1.2	7.1 ± 5.3
50 cm soil water	1.6 ± 1.7	1.3 ± 1.2	36.1 ± 21.7
75 cm soil water	1.6 ± 1.7	1.5 ± 1.6	11.6 ± 8.7
100 cm soil water	0.9 ± 1.1	1.6 ± 1.7	8.4 ± 6.3
Rainwater	—	93.7 ± 0.7	—

Table 2.
Contribution of different depths of soil water and rainwater to the water source of Pinus tabuliformis in hilly gully area (%, mean ± SD).

Water source	May 12	Jul. 15	Sep. 23
10 cm soil water	91.6 ± 3.3	7.4 ± 6.4	30.4 ± 10.4
25 cm soil water	5.7 ± 4.4	10.3 ± 8.8	27.1 ± 21.1
50 cm soil water	1.2 ± 1.2	21.1 ± 17.6	15.9 ± 12.6
75 cm soil water	0.7 ± 0.9	11.4 ± 9.7	14.2 ± 11.3
100 cm soil water	0.7 ± 0.9	13.1 ± 11.1	12.4 ± 9.9
Rainwater	—	36.8 ± 9.5	—

Table 3.
Contribution of different depths of soil water and rainwater to the water source of Armeniaca sibirica in hilly gully area (%, mean ± SD).

Moreover, the 18-yr *Robinia pseudoacacia* were sensitive to precipitation variation and used more deep soil water in a drier year, whereas the 30-yr *R. pseudoacacia* always used middle and deep soil water in wetter or drier year in the central region of Loess Plateau [27]. Therefore, woody species which could use deep soil water as a reliable water source during drought will have advantage in semi-arid region.

3. Water source of three shrubs in sandy land

3.1 Three sand binding shrubs in Mu Us Sandy Land

This study was conducted in Ordos Sandy Land Ecological Station, Institute of Botany, Chinese Academy of Forestry, which located in Ejin Holo Banner, Ordos

City, Inner Mongolia. The banner has a temperate continental climate, which mean air temperature varies from 5.0°C to 8.5°C, mean annual precipitation is 350 mm, potential evapotranspiration is 2300 mm and forest free days are 136 d [28]. The natural vegetation is sandy land dominated by *Sabina vulgaris*, *Artemisia ordosica*, *Salix psammophila* and *Caragana intermedia*, and steppe dominated by *Stipa bungeana* and *Iris lactea* var. *chinensis*.

Sabina vulgaris is an evergreen shrub with height of 0.3–1.0 m. It is a drought resistant species and widely used as sand binding and soil and water conservation shrub. It mainly inhabits on rocky slope, mixed coniferous and broad-leaf forest or sand dune, which distributes in Tianshan Mountain, Altai Mountain, Helanshan Mountain, Inner Mongolia, Northeast Qinghai, Gansu and North Shaanxi [22].

Artemisia ordosica is a small shrub with mean height of 0.5–1.0 m. It is resistant to sand burial and was widely used as a good sand-binding plant with air seeding. It mainly inhabits in moving, semi-fixed and fixed sand dunes in desert and slope in steppe, which distributes in Inner Mongolia, North Hebei and North Shanxi [29].

Salix psammophila (sand willow) is a shrub with mean height of 3–4 m. It is resistant to wind and sand burial, which was widely used as a sand-binding and afforestation species. It mainly distributes in Shaanxi, Inner Mongolia, Ningxia and Shanxi [30].

In the study site, the mean height of *S. vulgaris*, *A. ordosica* and *S. psammophila* were 0.78, 0.92 and 3.33 m, respectively. The semi-shrub *Hedysarum fruticosum* var. *laeve* was the associated species in *A. ordosica* and *S. psammophila* community. The main water source of four shrubs was measured by comparing their xylem water with different water sources, e.g. soil water in 10, 25, 50, 75, 100 and 150 cm for *A. ordosica*, whereas soil water in 10, 25, 50, 100, 150 and 200 cm for other two shrubs. The sample and extraction of soil water and xylem water was shown in 2.1. Groundwater was not sampled because shallow groundwater was depleted after coal mining in 2012. Deep groundwater table was 70 m at the study site, which is unavailable for vegetation. Meteorological data was obtained from Ordos Ecological Station. The contribution of different water sources to their total water use was analyzed by Iso-source 1.3.1 software [24].

3.2 Precipitation in Ordos Ecological Station in the growing season of 2018

The total precipitation of Ordos Ecological Station was 367.00 mm in the growing season of 2018 (**Figure 3**), which was slightly higher than the annual mean precipitation of 350 mm. The maximal daily precipitation was 34.4 mm and occurred on August 30, the next value was 32.2 mm on July 19, and 31.8 mm on July 16. The monthly precipitation were 25.6, 47.6, 11.4, 144.2, 92.2 and 46.0 mm from April to September.

3.3 Stable oxygen isotope of xylem water of four shrubs, soil water and rainwater in sandy land

On May 13, stable oxygen isotope ratio of *S. vulgaris* xylem water were closer to soil water in 10–25 cm, stable oxygen isotope ratio of *A. ordosica* and *H. fruticosum* var. *laeve* xylem water were closer to soil water in 10–150 cm, whereas stable oxygen isotope ratio of *S. psammophila* and *H. fruticosum* var. *laeve* xylem water was closer to soil water in 10 cm and 50–200 cm. On July 13, stable oxygen isotope ratio of 10 cm soil water was closer to rainwater on July 11. Stable oxygen isotope ratio of *S. vulgaris* was closer to soil water in 25 cm and 100–200 cm, stable oxygen isotope ratio of *A. ordosica* and *H. fruticosum* var. *laeve* xylem water were closer to soil water in 10 cm and 150 cm, whereas stable oxygen isotope ratio of *S. psammophila*

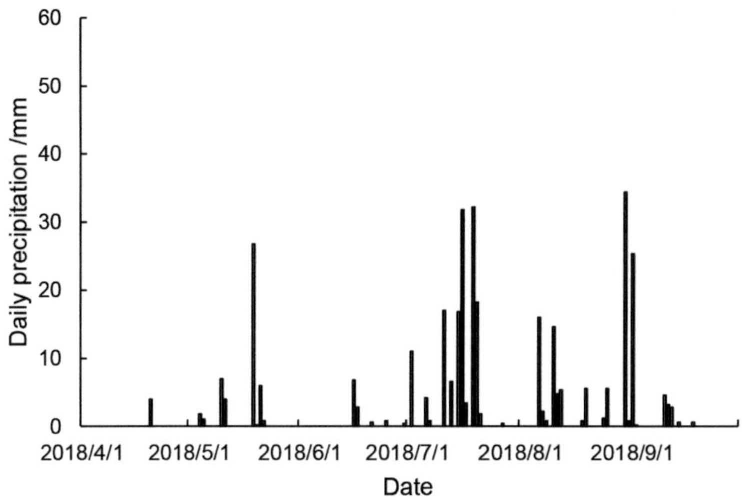

Figure 3.
Daily precipitation at Ordos Ecological Station from April to September of 2018.

and *H. fruticosum* var. *laeve* xylem water was closer to soil water in 10–25 cm and 100–200 cm. On September 22, stable oxygen isotope ratio of 10 cm soil water was closer to rainwater on September 11. Stable oxygen isotope ratio of *S. vulgaris* xylem water was closer to soil water in 25–200 cm, stable oxygen isotope ratio of *A. ordosica* and *H. fruticosum* var. *laeve* xylem water were closer to soil water in 10–150 cm, whereas stable oxygen isotope ratio of *S. psammophila* and *H. fruticosum* var. *laeve* xylem water were closer to soil water in 25–200 cm (**Figure 4**).

3.4 Contribution of different depth of soil water to the water source of four shrubs in sandy land

Iso-source analysis showed that *S. vulgaris* mainly used 25 cm soil water on May 13, which accounted for 78.5% of its total water source. On July 17, it mainly used 10–25 cm and 100–200 cm soil water and rainwater, which accounted for 50.8% and 40.7% of its total water source. On September 22, it mainly used 10–25 cm and 100–200 cm soil water, which accounted for 51.2% and 39.8% of its total water source (**Table 4**).

Iso-source analysis showed that *A. ordosica* and *H. fruticosum* var. *laeve* mainly used 10 cm soil water on May 13, which accounted for 72.9% and 66.5% of their total water source, respectively. On July 17, they mainly used 10 cm and 150 cm soil water, which accounted for 65.8% and 54.9% of their total water source, respectively. On September 22, they mainly used 10–25 cm soil water, which accounted for 46.1% and 49.0% of their total water source, respectively (**Table 5**).

Iso-source analysis showed that *S. psammophila* and *H. fruticosum* var. *laeve* mainly used 10–25 cm soil water on May 13, which accounted for 59.0% and 37.9% of their total water source. On July 17, they mainly used 50–200 cm soil water and 10–25 cm and 100–200 cm soil water, which accounted for 71.0% and 91.8% of their total water source, respectively. On September 22, they mainly used 10–100 cm soil water, which accounted for 73.8% and 72.5% of their total water source, respectively (**Table 6**).

Four shrubs in sandy land have a resource-dependent water use strategy, e.g. used different depths of soil water based on their availability in the growing season. *Sabina vulgaris* mainly used 25 cm shallow soil water in spring; however, it mainly

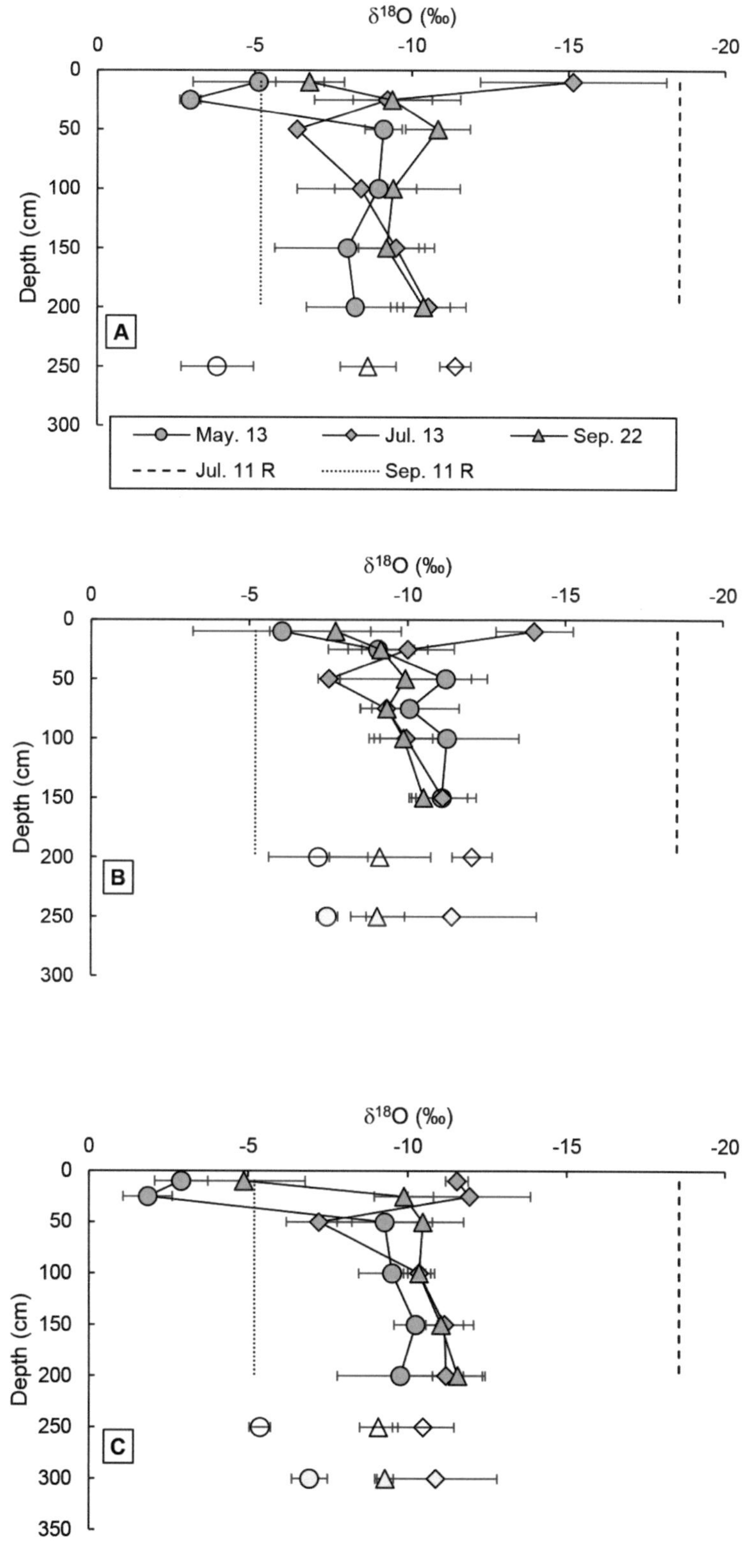

Figure 4.
Stable oxygen isotope ratio of xylem water, soil water and rainwater of Sabina vulgaris (A), Artemisia ordosica (B) and Salix psammophila (C) in sandy land. Dark symbol is soil water, white symbol is xylem water of three shrubs, and gray symbol is xylem water of Hedysarum fruticosum var. laeve.

Water source	May 13	Jul. 17	Sep. 22
10 cm soil water	8.9 ± 7.7	38.1 ± 5.5	36.8 ± 5.5
25 cm soil water	78.5 ± 4.9	12.7 ± 10.7	14.4 ± 12.1
50 cm soil water	2.8 ± 2.7	8.4 ± 7.2	9.1 ± 7.8
100 cm soil water	2.9 ± 2.8	11.0 ± 9.4	14.2 ± 12.0
150 cm soil water	3.6 ± 3.3	13.3 ± 11.3	15.4 ± 13.0
200 cm soil water	3.4 ± 3.2	16.4 ± 13.8	10.2 ± 8.8

Table 4.
Contribution of different depths of soil water and rainwater to the water source of Sabina vulgaris in sandy land (%, mean ± SD).

Water source	May 13		Jul. 17		Sep. 22	
	Ao	Hf	Ao	Hf	Ao	Hf
10 cm soil water	72.9 ± 3.1	66.5 ± 3.6	51.9 ± 5.5	36.6 ± 7.0	26.0 ± 7.9	30.0 ± 7.5
25 cm soil water	7.9 ± 6.9	9.7 ± 8.4	9.9 ± 8.5	13.1 ± 11.1	20.1 ± 16.8	19.0 ± 16.0
50 cm soil water	4.4 ± 4.0	5.5 ± 4.9	5.9 ± 5.3	7.9 ± 6.8	12.9 ± 11.0	12.2 ± 10.4
75 cm soil water	5.8 ± 5.1	7.2 ± 6.3	8.4 ± 7.3	11.2 ± 9.5	17.6 ± 14.8	16.7 ± 14.0
100 cm soil water	4.4 ± 4.0	5.5 ± 4.9	9.8 ± 8.4	13.0 ± 11.0	13.3 ± 11.2	12.5 ± 10.7
150 cm soil water	4.5 ± 4.1	5.7 ± 5.0	13.9 ± 11.8	18.3 ± 15.4	10.1 ± 8.7	9.6 ± 8.2

Table 5.
Contribution of different depths of soil water and rainwater to the water source of Artemisia ordosica (Ao) and Hedysarum fruticosum var. laeve (Hf) in sandy land (%, mean ± SD).

Water source	May 13		Jul. 17		Sep. 22	
	Sp	Hf	Sp	Hf	Sp	Hf
10 cm soil water	28.6 ± 18.5	19.5 ± 12.1	15.2 ± 12.8	18.0 ± 14.6	26.4 ± 3.2	22.7 ± 3.4
25 cm soil water	30.4 ± 16.0	18.4 ± 10.5	13.8 ± 11.7	16.6 ± 13.3	16.9 ± 14.2	17.8 ± 14.9
50 cm soil water	10.8 ± 8.8	16.2 ± 13.0	16.9 ± 5.1	8.2 ± 4.3	15.1 ± 12.7	15.8 ± 13.3
100 cm soil water	10.5 ± 8.5	15.8 ± 12.7	20.9 ± 17.4	18.6 ± 14.1	15.4 ± 13.0	16.2 ± 13.6
150 cm soil water	9.6 ± 7.8	14.7 ± 11.7	16.7 ± 14.0	19.4 ± 16.0	13.7 ± 11.6	14.3 ± 12.1
200 cm soil water	10.1 ± 8.3	15.4 ± 12.3	16.5 ± 13.8	19.2 ± 15.8	12.6 ± 10.7	13.2 ± 11.2

Table 6.
Contribution of different depths of soil water and rainwater to the water source of and Salix psammophila (Sp) and Hedysarum fruticosum var. laeve (Hf) in sandy land (%, mean ± SD).

used both of 10–25 cm shallow and 100–200 cm deep soil water in summer and autumn. *Artemisia ordosica* and the accompany plant *H. fruticosum* var. *laeve* always used the same water source, e.g. 10 cm surface soil water in spring, 10 cm surface soil water and 150 cm deep soil water in summer, and 10–150 cm soil water in autumn. *Salix psammophila* mainly used 10–25 cm shallow soil water whereas *H. fruticosum* var. *leave* mainly used 50–200 cm soil water in spring. However, they both mainly used 10–25 cm shallow soil water and 100–200 cm deep soil water in summer, and 25–200 cm soil water in autumn. Soil water recharged by rainwater is the main water source for vegetation in sandy land since shallow groundwater was depleted by coal mining. However, in the habitat with shallow groundwater in the sandy land, woody species still could use groundwater. For example, *Salix matsudana* and *S. vulgaris* mainly used deep soil water and groundwater, whereas *A. ordosica* mainly used groundwater in Mu Us Sandy Land. The groundwater table were 1.0 m for *S. matsudana* in interdune, 1.5 m and 1.3 m for *S. vulgaris* and *A. ordosica* on sand dune, respectively [5].

The water source of the company shrub *H. fruticosum* var. *leave* were similar to the dominant shrub *A. ordosica* or *S. psammophila*, which indicated water competition between them. Water source of shrubs was closely related to their root type in sandy land, especially fine root. The root length of *H. fruticosum* var. *laeve* was 80 cm, with the biomass concentrated within 40 cm [31]. The distribution of its root system is partly overlap with two dominant shrubs. The root depth of *A. ordosica* was 200 cm, with the fine root concentrated within 40 cm [32]. The root system of *S. psammophila* was as deep as 150 cm, with fine root concentrated within 50 cm [33]. Similar phenomenon occurred in other sandy land vegetation. For example, *Salix gordejevii* mainly used soil water, whereas *Artemisia halodendron* mainly used 10–150 cm soil water in Horqin Sandy Land [11]. Their root biomass concentrated within 40 cm in the mixed community [34]. Therefore, the distribution of deep-rooted species should be arranged appropriately to avoid excessive water competition in the restoration of degraded vegetation in sandy land. It was suggested to keep reasonable afforestation density in the ecological engineering in Mu Us Sandy Land and other area of semi-arid region.

4. Conclusion

Soil water recharged by rainwater is the main water source for dominant species on Ordos Plateau. Six woody species have resource-dependent water use strategy, which is an adaptive advantage to the semi-arid climate. On loess slope of eastern Ordos Plateau, *H. rhmnoides, P. tabuliformis* and *A. sibirica* selected different water source based on their availability in the growing season, including shallow soil water, deep soil water and rainwater. In Sandy Land of southern Ordos Plateau, *S. vulgaris*, *A. ordosica*, *S. psammophila* and the company semi-shrub *H. fruticosum* var. laeve used different depths of soil water based on their availability in the growing season. However, there were water competition between the company semi-shrub and two dominant shrubs. Therefore, it was suggested to avoid vegetation degradation resulted from excessive water competition by appropriate distribution and afforestation density in semi-arid region.

Acknowledgements

The author thanks Ordos Sandy Land Ecological Station for supporting meteorological data.

This study was supported by National Key Research & Development Program of China (2017YFC05040450202) from Ministry of Science and Technology of the People's Republic of China.

Conflict of interest

The author declares no conflict of interest.

Author details

Yajuan Zhu
Institute of Desertification Studies, Chinese Academy of Forestry, Beijing, China

*Address all correspondence to: zhuyj@caf.ac.cn

References

[1] Zhang X S. Principles and optimal models for development of Maowusu Sandy Grassland. Acta Phytoecologica Sinica. 1994;18(1):1-16.

[2] Lin G. Stable Isotope Ecology. Beijing: Higher Education Press; 2013. 492p.

[3] Cheng X L, An S Q, Li B, Chen J Q, Lin G H, Liu Y H, Luo Y Q, Liu S R. Summer rain pulse size and rainwater uptake by three dominant desert plants in a desertified grassland ecosystem in northwestern China. Plant Ecology. 2006;184(1):1-12. DOI: 10.1007/s11258-005-9047-6

[4] Yang H, Auerswald K, Bai Y F, Han X G. Complementarity in water sources among dominant species in typical steppe ecosystems of Inner Mongolia, China. Plant and Soil. 2011;340 (1/2):303-313. DOI: 10.1007/S11104-010-0307-4

[5] Ohte N, Koba K, Yoshikawa K, Sugimoto A, Matsuo N, Kabeya N, Wang L H. Water utilization of natural and planted trees in the semiarid desert of Inner Mongolia, China. Ecological Applications. 2003;13(2):337-351.

[6] Wei Y F, Fang J, Liu S, Zhao X Y, Li S G. Stable isotopic observation of water use sources of *Pinus sylvestris* var. *mongolica* in Horqin Sandy Land, China. Trees. 2013;27(5):1249-1260. DOI: 10.1007/s00468-013-0873-1

[7] Su H, Li Y G, Liu W, Xu H, Sun O J. Changes in water use with growth in *Ulmus pumila* in semiarid sandy land of northern China. Trees. 2014;28(1):41-52. DOI: 10.1007/s00468-013-0928-3

[8] Jian J, Jia D, Guo S, Qian L. Water sources in growing season of *Salix gordejevii* in the Otindag Sandy Land traced by stable D isotope in 2014. Arid Land Research. 2017;34(2):350-355. DOI:10.13866/j.azr.2017.02.15

[9] Zhu Y J, Wang G J, Li R Q. Seasonal dynamics of water use strategy of two *Salix shrubs* in alpine sandy land, Tibetan Plateau. PLoS One. 2016;11(5):e0156586. DOI:10.1371/journal.pone.0156586

[10] Zhu Y, Qi K, Pang Z. Water source of *Salix cheilophila* plantation in alpine sandy land in summer. Journal of Nanjing Forestry University (Natural Sciences Edition). 2019;43(1):91-97. DOI: 10.3969/j.issn.1000-2006.201806036

[11] Liu B, Liu Z, Qian J, Alamusa, Zhang F, Peng X. Water sources of dominant sand-binding plants in dry season in southern Horqin Sandy Land, China. Chinese Journal of Applied Ecology. 2017;28(7):2093-2101. DOI: 10.13287/j.1001-9332.201707.030

[12] Song L, Zhu J, Li M, Yu Z. Water utilization of *Pinus sylvestris* var. *mongolica* in a sparse wood grassland in the semiarid sandy region of Northeast China. Trees. 2014;28(4):971-982. DOI: 10.1007/s00468-014-1010-5

[13] Chen X, Chen Z, Zhao Y. The determination of ecotone and the characteristics of biome on Ordos Plateau. Acta Phytoecologica Sinica. 1998;22(4):312-318.

[14] Wang R, Yan F, Wang Y. Vegetation growth status and topographic effects in the Pisha Sandstone Area of China. Remote Sensing. 2020;12. DOI: 10.3390/rs12172759

[15] Bai Z, Cui J, Ding X. Desertification and its driving factors in the Ordos Plateau, from 1986 to 2015. Arid Zone Research. 2020;37(3):749-756. DOI: 10.13866/j.azr.2020.03.24

[16] Yao W, Wu Z, Liu H, Xiao P, Yang C. Experimental research on the anti-erosion and vegetation promotion for sandstone region in the Yellow River Basin. Yellow River. 2015;37(1):6-10.

[17] Cao Y, Pang Y, Jia X. Vegetation growth in Mu Us sandy land from 2001 to 2016. Bulletin of Soil and Water Conservation. 2019;39(2):29-37.

[18] Chen Y, Dong M. Quantitative analysis of landscape conditions of the desertified sandy grassland in Ordos Plateau. Environmental Science. 2002;23(1):87-91.

[19] Wang J, Liu L, Jia K, Tian L. Spatiotemporal variation of vegetation phenology and its affecting factors in the Mu Us Sandy Land. Journal of Desert Research. 2015;35(3):624-631. DOI: 10.7522/j.issn.1000-694X.2015.00021

[20] Liu Z, Zhang J, Zhang E. Model of returning crop plots to forestry and grassland in hilly and gully area of Jungar Banner, Inner Mongolia. Journal of Desert Research. 2002;22(5):506-509.

[21] Fang W, Chang C, editors. Flora Reipublicae Popularis Sinicae, Tomus 52 (2): Dicotyledoneae, Elaeagnaceae, Lecythidaceae, Lythraceae, Rhizophoraceae, Sonneratiaceae, Nyssaceae, Crypteroniaceae, Alangiaceae, Punicaceae. Beijing: Science Press; 1978. 542p.

[22] Cheng W, Fu L, editors. Flora Reipublicae Popularis Sinicae, Tomus 7: Gymnospermae. Beijing: Science Press; 1978. 542p.

[23] Yu T, editor. Flora Reipublicae Popularis Sinicae, Tomus 38: Dicotyledoneae, Rosaceae (3), Prunoideae, Connaraceae. Beijing: Science Press; 1986. 167p.

[24] Phillips D L, Gregg J W. Source partitioning using stable isotopes: coping with too many sources. Oecologia. 2003;136(2):261-269.

[25] Yang G, Wang A, Wang L. Water source and water use efficiency of two typical shrubs in different seasons in Liudaogou Watershed. Acta Botanica Boreali-Occidentalia Sinica. 2018;38(1):140-149. DOI: 10.7606/j.issn.1000-4025.2018.01.0140

[26] Wang J, Fu B, Wang L, Lu N, Li J. Water use characteristics of the common tree species in different plantation types in the Loess Plateau of China. Agricultural and Forest Meteorology. 2020;288-289:108020. DOI: 10.1016/j.agrformet.2020.108020

[27] Wang J, Fu B, Jiao L, Lu N, Li J, Chen W, Wang L. Age-related water use characteristics of *Robinia pseudoacacia* on the Loess Plateau. Agricultural and Forest Meteorology. 2021;301-302:108344. DOI: 10.1016/j.agrformet.2021.108344

[28] Ye X H, Liu Z L, Zhang S D, Gao S Q, Liu G F, Cui Q G, Du J, Huang Z Y, Cornelissen J H C. Experimental sand burial and precipitation enhancement alter plant and soil carbon allocation in a semi-arid steppe in north China. Science of the total Environment. 2019;651:3099-3106. DOI: 10.1016/j.scitotenv.2018.10.208

[29] Ling Y, Ling Y, editors. Flora Reipublicae Popularis Sinicae, Tomus 76(2): Dicotyledoneae, Compositae (3), Anthemideae (2). Beijing: Science Press; 1991. 333p.

[30] Wang C, Fang C, editors. Flora Reipublicae Popularis Sinicae, Tomus 20(2): Angiospermae, Dicotyledoneae, Salicaceae. Beijing: Science Press; 1984. 423p.

[31] Zhang L, Wang X, Hong G, Wu Y, Li Z, Hai L, Wang P, Gao X, Yang J. Root distribution characteristics of *Hedysarum laeve* with different aerial

seeding years in Mu Us Sandy Land. Chinese Journal of Ecology. 2017;36(1):29-34. DOI:10.13292/j.1000-4890.201701.001

[32] Zhang J, Han H, Lei Y, Yang W, Li Y, Yang D, Zhao X. Correlations between distribution characteristics of *Artemisia ordosica* root system and soil moisture under different fixation stage of sand dune. Journal of Southwest Forestry University. 2012;32(6):1-5. DOI:10.3969/j.issn.2095-1914.2012.06.001

[33] Liu J, He X, Bao H, Zhou C. Distribution of fine roots of *Salix psammophila* and its relationship with soil moisture in Mu Us Sandland. Journal of Desert Research. 2010;30(6):1362-1366.

[34] Huang G, Zhao X, Zhao Y, Su Y. Root distribution of two typical shrubs in single or mixture circumstance in Horqin Sandy Land. Journal of Desert Research. 2007;27(2):239-243.

Chapter 6

Characteristic on the Stability of *Haloxylon ammodendron* Plantation in the Southern Fringe of Gurbantunggut Desert, Northwest China

Qinghong Luo, Qimin Chen, Miao He and Na Li

Abstract

Using chronosequence theory and method, the characteristics of vegetation-soil coupling and structure stability of *Haloxylon ammodendron* plantations in the southern fringe of Gurbantunggut Desert were analyzed. The results showed, the canopy storey of *H. ammodendron* plantation experienced three stages, rapid growth (the age of 7 to 20), then slow growth (the age of 20 to 28) and last decline (over the age of 28). The best natural regeneration started from 17-yr-old plantation. Vegetation-soil system coupling degree (C) and coupling coordinative degree (D) of plantations with different age were not one-to-one correspondence. The system of *H. ammodendron* plantations always stayed in disorder recession, vegetation and soil were prone to loss type during the process of sand-fixation. Five principal components evaluated that the first rank was 42-yr-old plantation. It was inferred that the trend of the vegetation and soil system was from senescence to harmonious development. So the trend of coordinated development between vegetation and soil would be promoted, if the artificial tending and management measures strengthened.

Keywords: *Haloxylon ammodendron* plantation, structure dynamic, survive and death curve, vegetation-soil coupling degree, coupling coordinative degree

1. Introduction

Structure and function are two vital aspects to determine and measure the stability of forests community [1]. Population structure, which reflects the distribution of individuals within a population, may provide insights into past and present regeneration [2, 3]. Life table and survival curve are the main methods when analyzing population structure. By using life table analysis, it was discovered that *H. ammodendron* seedling had two rapid drop stages of survival rate and high mortality rate in natural forest [4]. Meanwhile, by using survival curve analysis, *H. ammodendron* population was suggested closing to progressive or stable status in some suitable habitats [5, 6]. This, however, has been unknown in the plantation which had similar inhabits but different growth stage.

Plantation has been applied in ecological restoration worldwide [7]. A key issue for plantation is that population regeneration is often challenged by seedling establishment in many plantation ecosystems, which is critical to the stability of plantation species as well as to the plantation sustainability [8]. For *H. ammodendron*, failure in regeneration resulted in the population degradation occurred in old stand in the Hexi Corridor Desert [9, 10], suggesting that the regeneration barrier possibly occurs in *H. ammodendron* plantation. This, however, has not been assessed in the biggest *H. ammodendron* plantation developed in the Gurbantunggut Desert edges.

The wind prevention and sand fixation forest ecosystem, which is a kind of relatively heterogeneous open system, and flow of the material, energy and information constantly happens among the subsystems, thereinto, the two major subsystems, vegetation and soil formed an interdependent and mutually restrictive relationship between them [11]. Exploring the relationship between vegetation and soil has significant meaning in understanding the process and effect of sand fixation vegetation, and further putting forward corresponding optimal management measures [12]. In the ecological process of settlement, growth and succession of artificial *H. ammodendron* population, the accumulation and distribution of soil moisture and nutrient will change. In turn, that will affect the characteristics of vegetation. Therefore, such mutual restriction relationship between vegetation and soil will definitely change according to the spatial and temporal scale [13].

The Gurbantunggut Desert in northwestern China, the third largest desert in the world, is sensitive to climate change and human activities [14]. The desert vegetation is dominated by the *H. ammodendron*, which is greatly tolerant to drought, wind erosion, sand burial and other stress factors [15–18]. Therefore, *H. ammodendron* has been widely used in vegetation restoration and sand-fixation engineering [19]. In the edge of the Gurbantunggut Desert, i.e., the desert-oasis ecotone, *H. ammodendron* forest play an important role in preventing dune movement toward oasis. However, large area of *H. ammodendron* forest was suffered deforestation and overgrazing during the 1970s and 1980s, which leads to severe desertification in this area [20]. Since then *H. ammodendron* plantation has been extensively developed along the edge of the Gurbantunggut Desert. In recent decades, these *ammodendron* plantations have gradually developed into the mature stage, but, some of them tend to decline, which directly threatened the stability of the artificial population system and the normal exertion of ecological functions in this area.

This chapter, the change characteristics of the canopy storey and the regeneration storey structures of *H. ammodendron* plantations at different chronosequence stages, as well as the growth and death dynamic development law of populations ware studied. Meanwhile, the feedback relationship between vegetation and soil were analyzed. Further, the current situation of growth and development of *H. ammodendron* population in the edge of the Gurbantunggut Desert was evaluated accurately and comprehensively. Then, we explained why the *H. ammodendron* population is stable, developing or declining, and whether *H. ammodendron* plantation has the ability of sustainable natural regeneration. On this basis, we put forward the scientific suggestion for the vegetation construction and management of the degraded artificial forest ecosystem in this region. We hope this research could provide basic data and scientific evidence for further research on the vegetation restoration and reconstruction of desert ecosystem in arid region.

2. Characteristics of vegetation and soil system of *H. ammodendron* plantation

2.1 Basal characteristic of *H. ammodendron* plantations in study site

In Mosuowan Reclamation Area, Corps 150, southern edge of the Gurbantunggut Desert, *H. ammodendron* plantations were chosen for this study, which are constructed on moving sandy land within an area of about 57.61 km^2. Thus soil and climate is similar among these plantations. All plantations were established by transplanting 1-yr-old *H. ammodendron* seedlings in regular spacing (**Table 1**). Thus *H. ammodendron* individuals could be classified into two main storeys, i.e. transplanted individuals (canopy storey) and seedlings including those naturally regenerated between transplanted ones (regeneration storey). None of the plantations were subjected to irrigation, pruning, thinning or any other silvicultural practices after transplantation. In addition, no incident of insect, rodents and fire was recorded since 1981. Except *H. ammodendron* seedlings, understory plants were dominated by *Ceratocarpus arenarius* and *Salsola ruthenica*, with a low total cover less than 8% (**Table 1, Figure 1**).

In August, when soil water content of the study site was relatively stable and close to the average level in a year [21], vegetation survey was carried out in nine *H. ammodendron* plantations. In each plantation, three 20 m × 30 m sample plots were established, with each at least 100 m away from the others. Within each sample plots, the numbers of survived and dead individuals for canopy storey was recorded (**Table 1**). Meanwhile, height, basal stem diameter and crown width (north–south and east–west) of survived individuals were measured one by one. Then the crown projected area, above-ground biomass and canopy density of major storey were calculated according to Phillips and Macmahon's ellipse formula [22] and Song's Eq. [23], respectively.

2.2 Characteristics of canopy storey in six *H. ammodendron* plantations at different Chronosequence stage

Vegetation survey was carried out in six *H. ammodendron* plantations, which having capacity of normal natural regeneration. According to Rundel's method [24], combining with the characteristic of slow growth and false ring of H. ammodendron [25], the structure of canopy storey was analyzed by quantity and size dynamic of tree height and basal stem diameter. Tree height class was divided into 0 ~ 0.5 m, 0.5 ~ 1.0 m, 1.0 ~ 1.5 m, 1.5 ~ 2.0 m, 2.0 ~ 2.5 m, 2.5 ~ 3.0 m, 3.0 ~ 3.5 m, 3.5 ~ 4.0 m, 4.0 ~ 4.5 m, 4.5 ~ 5.0 m, 5.0 ~ 5.5 m and 5.5 ~ 6.0 m. Basal stem diameter was divided into 0 ~ 1 cm, 1 ~ 2 cm, 2 ~ 4 cm, 4 ~ 6 cm, 6 ~ 8 cm, 8 ~ 10 cm, 10 ~ 12 cm, 12 ~ 14 cm, 14 ~ 16 cm, 16 ~ 18 cm and 18 ~ 22 cm. Structural characteristics of main layer were analyzed by quantity and size.

As shown in **Figures 2** and **3**, in the growth process of *H. ammodendron*, the distribution centers of height class and basal stem diameter class of the canopy storey

Chronosequence Stage (yr)	Location (N, E)	Altitude (m)	Planting space (m)	Preserving rate (%)
7	45°05′4.4″, 85°59′26.9″	330	1.5 × 2	99.3
12	45°01′56.8″, 86°08′22.2″	332	2 × 3	98.2
15	44°39′35.4″, 86°20′12.4″	344	2 × 2	91.4
17	45°08′5.4″, 85°59′26.0″	319	1 × 5	100
20	45°02′48.8″, 86°08′48.7″	328	3 × 3	97.2
23	44°34′29.1″, 83°18′49.4″	310	2 × 2	81.9
28	45°07′48.8″, 85°58′35.0″	314	3 × 4	95.4
33	45°10′33.5″, 85°55′35.1″	309	1 × 5	93.4
42	44°01′56.8″, 86°08′22.3″	320	1 × 5	91.7

Table 1.
The sample plots information of H. ammodendron *plantations at different chronosequence stage.*

Figure 1.
The original appearance (A), land preparation (B), the H. ammodendron *plantation with age of 7 (C), 17 (D), 28 (E), 42 (F).*

moved to a higher section gradually, and the structure of size class showed a trend of continuous differentiation first and then concentration. It indicated that the intra-specific competition among individuals was intensifying, and the population structure tended to be more and more complex, which was conducive to the rational use of spatial resources. The tree height and basal stem diameter class of canopy storey showed obvious differentiation in the plantations with the age of 17 and 20. It indicated that, after the competition and self-thinning, some individuals having competitive advantage had entered in the rapid growth stage, and most individuals could grow to the canopy storey (HT 2.5–3.0 m, BSD 8-10 cm). After the age of 28, *H. ammodendron* plantation would start to decline, when branch aging and withering appeared, and the natural death rate of *H. ammodendron* in canopy storey would reach 5% ~ 8%.

2.3 Characteristics of regeneration layer in six *H. ammodendron* populations at different chronosequence stage

2.3.1 Characteristics of seedling with different grade

Following the method of the Blackman [26] and Forest resources planning survey [27], seedlings in regeneration storey were divided into three grades, namely grade I

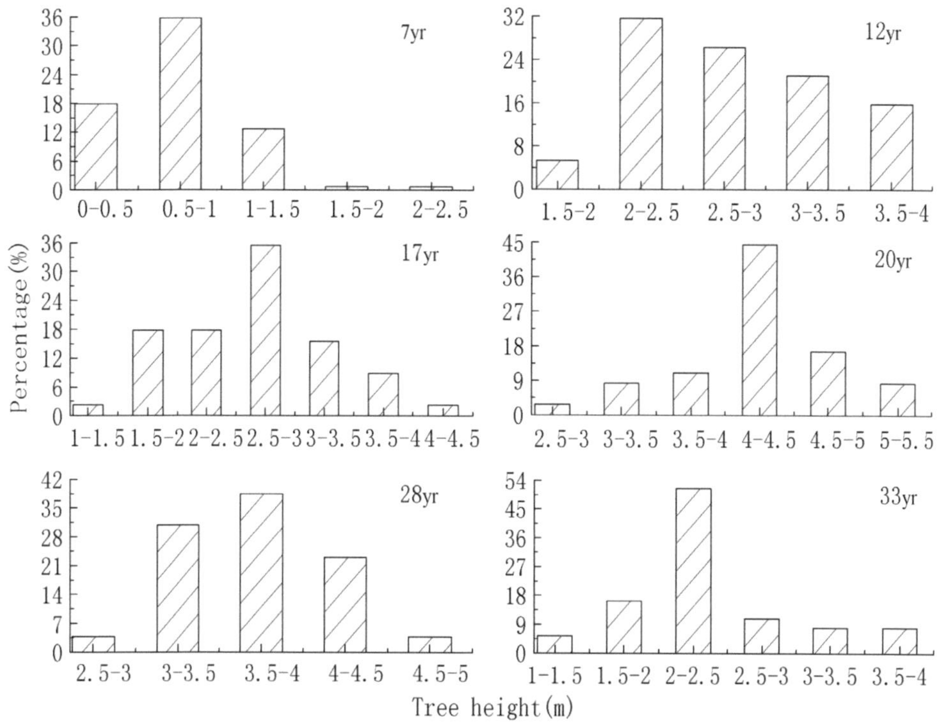

Figure 2.
Dynamitic distribution of H. ammodendron *height (HT) in canopy storey at different chronosequence stage.*

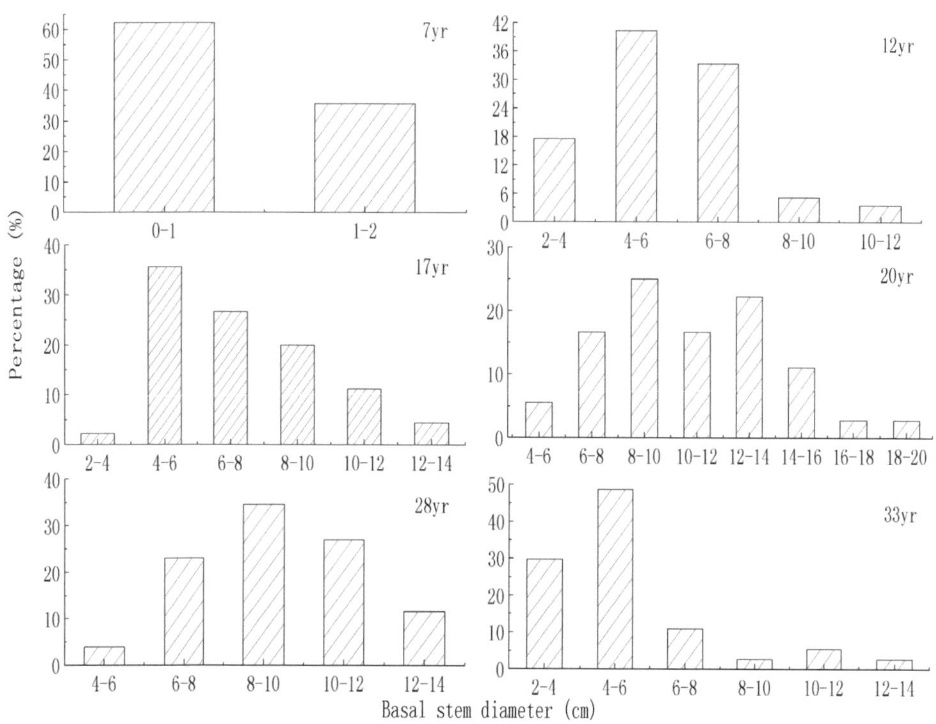

Figure 3.
Dynamitic distribution of H. ammodendron *basal stem diameter (BSD) in canopy layer at different chronosequence stage.*

seedling was HT ≤30 cm, grade II seedling was 31 ≤ HT <50 cm, and grade III seedling was HT ≥50 cm. respectively. The survival numbers of each grade seedlings were recorded respectively according to the following criteria. The grade I seedling status was considered 'Good', 'Medium' and 'Bad', if their number was >5000, 3000 ~ 4999 and < 2999. Grade II seedling status was considered 'Good', 'Medium' and 'Bad', if their number was >3000, 1000 ~ 2999 and < 999. Grade III seedling status was considered 'Good', 'Medium' and 'Bad', if their number was ≥5000, 500 ~ 4990 and < 500.

It can be seen from **Table 2** and 7-yr-old plantation already had the weaker ability of regeneration. In the plantation age of 7, 12, 17, 28, and 33, grade I seedlings all accounted greater proportion. Especially in 28-yr-old plantation, 67.3% of the total regenerated individuals (36,383 No.hm^{-2}) was grade I seedling, and grade III only accounted for 1.9%. Grade I seedlings were presented "bad" in 20-yr-old plantation. Grade II seedlings were "good" in the plantations with the age of 17 to 28. Grade III seedlings were "good" only in 17 and 33 yrs. old plantation. In contrast, there were not only abundant grade I seedlings but also larger proportion of grade III individuals (29.3%) in 17-yr-old plantation. For 33-yr-old plantation, although seedling density was only 9,433 No.hm^{-2}, the ratio of grade III reached 40.3%. The growth quality of grade III seedling was better than others apparently in 12-yr-old and 17-yr-old plantations. Especially in the 17-yr-old one, the average height and basal stem diameter of grade III seedling even reached 1.2 m and 1.9 cm, respectively (**Table 2**).

The quantity and growth quality of older seedlings is more important for the sustainable development of population [7]. As shown in **Figure 4**, the best growth status of grade III seedlings was in 17-yr-old plantation, average height and basal stem diameter reached 1.10 m and 1.91 cm respectively, and the maximum were 2.19 m and 3.89 cm, respectively. Height of grade III seedlings was more evenly in 28-yr-old plantation, which was mainly distributed between 0.51 ~ 1.02 m. For 33-yr-old plantation, the density and growth of the grade III seedlings were similar with the 28 yrs. old one.

2.3.2 Analysis on influencing factors of natural regeneration in H. ammodendron *plantation*

Results of the relational degree (Rd) between natural regeneration and its influencing factors shown in **Table 3**. Among six vegetation factors, above-ground biomass of the *H. ammodendron* individuals in canopy storey was closely related with density of seedlings in the regeneration storey (Rd = 0.77). But the density of grade III seedlings was mainly influenced by the density (Rd = 0.71) and age (Rd = 0.70) of the individuals in the canopy storey. Furthermore, density of seedlings had greatest Rd. with SHN (Rd = 0.87), as followed by SEP (Rd = 0.84) and SWC (Rd = 0.79). While the density, height and basal stem diameter of grade III

Grade	Chronosequence stage (yr)											
	7		12		17		20		28		33	
	Density	Rank	Density	Rank	Density	Rank	Density	Rank	Density	Rank	Density	Rank
I	100	B	4300	G	8733	G	783	M	24483	G	4117	G
II	683	B	1983	M	7200	G	3150	G	11200	G	1517	M
III	33	B	583	B	6600	G	1000	B	700	B	3800	M

Grade I: HT < 30 cm, Grade II: 31 ≤ HT < 50 cm, Grade III: HT ≥ 50 cm. The unite of density is No.hm^{-2}. 'G, M and B' in the table stand for 'good, medium and bad' respectively.

Table 2.
Quality and quantity of seedlings with three grades in H. ammodendron *plantation at different chronosequence stage.*

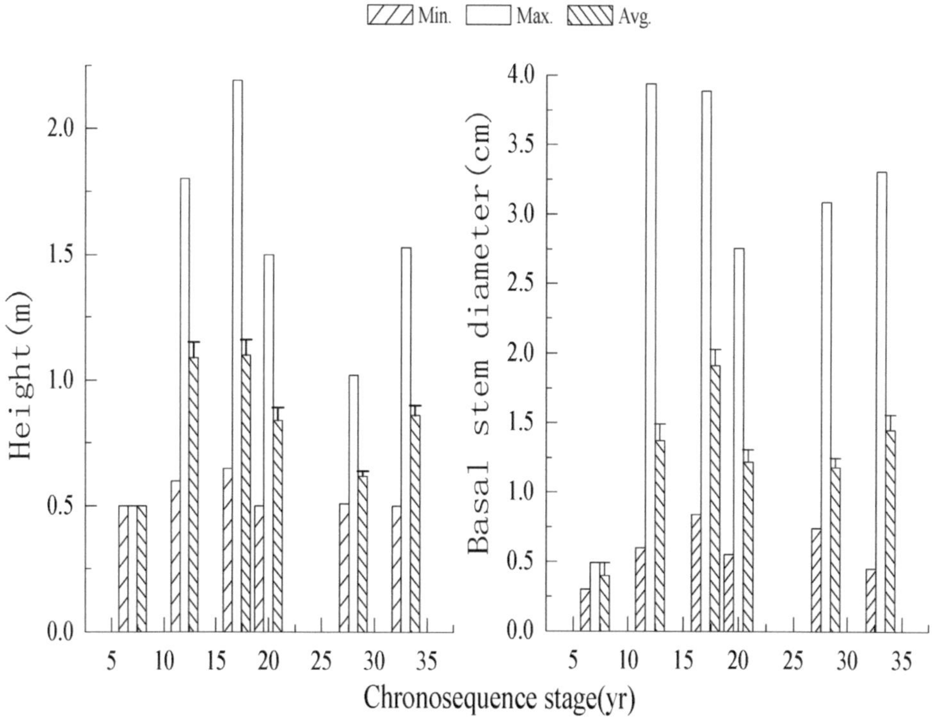

Figure 4.
Height and Basal Stem Diameter of grade III seedling (HT ≥ 50 cm) in H. ammodendron *plantation at chronosequence stage.*

Rd	X1j	X2j	X3j	X4j	X5j	X6j	X7j	X8j	X9j	X10j	X11j	X12j	X13j
X1i	0.68	0.72	0.62	0.68	0.72	0.77	0.79	0.75	0.76	0.66	0.87	0.84	0.73
X2i	0.70	0.71	0.62	0.64	0.65	0.68	0.73	0.65	0.63	0.78	0.77	0.72	0.64
X3i	0.69	0.70	0.75	0.70	0.69	0.63	0.82	0.74	0.81	0.87	0.72	0.67	0.72
X4i	0.74	0.67	0.71	0.78	0.71	0.66	0.76	0.65	0.76	0.94	0.71	0.70	0.77

X1i: density of seedling (No·hm^{-2}), X2i: density of grade III seedling (No·hm^{-2}), X3i: height of grade III seedling (m), X4i: basal stem diameter of grade III seedling (cm), X1j: plantation age (yr), X2j: planting density (No·hm^{-2}), X3j: density of canopy storey (%), X4j: height of regeneration storey (m), X5j: crown projected area of canopy storey (m2·tree^{-1}), X6j: Above-ground biomass of canopy storey (kg·tree^{-1}), X7j: SWC(%), X8j: pH, X9j: EC (ms·cm^{-1}), X10j: SOM (g·kg^{-1}), X11j: SHN (mg·kg^{-1}), X12j: SEP (mg·kg^{-1}), X13j: SAP (mg·kg^{-1}).

Table 3.
The relational degree (Rd) between characteristic of seedling and its related influencing factors.

seedlings were strongly influenced by SOM, Rd. was 0.78, 0.87 and 0.94 respectively, which explained that influence of SOM become more important than SHN for grade III seedlings. The soil pH of 8.1 to 8.6 had less influence on natural regeneration. In Minqin Desert, when soil pH was equal to or higher than 8.6, natural regeneration of *H. ammodendron* would be limited [28]. On the whole, soil environmental factors generally had greater influence on natural regeneration than that of vegetation factors of the canopy storey.

2.4 Population development characteristics of *H. ammodendron* plantation

Static Life Table was made according to relevant parameters calculated by actual measurment data of survival individual (a_x) in canopy storey and regeneration storey, the mutual relationship of parameters were as follow: $l_x = a_x/a_0 \times 1000$, $d_x = l_x - l_{x+1}$, $q_x = d_x/l_x \times 100\%$, $L_x = (l_x + l_{x+1})/2$, $T_x = \sum_x^{\infty} Lx$, $e_x = T_x/l_x$. Where a_x is the existing number of age x, l_x was standardized number, d_x was number of death interval, q_x was average survival rate per age stage, L_x was life between the interval, T_x was total life, e_x was mean expectation of life. When *H. ammodendron* individuals in canopy storey grew up to the age of 17, q_x rose obviously at age class above V, and the highest q_x appeared at age class below IV. In 7-yr-old plantation, the largest e_x was at age class I. And in the plantation with the age of 12, 17, 20, 28 and 33, the largest e_x presented at age class IV, II, VI, III and IV, respectively. It inferred that survival ability of *H. ammodendron* could be declined gradually with its growth (**Table 4**).

Variation tendency of survival and death curves were similar in the same plantation, and discrepancy still existed among six *H. ammodendron* plantations. The survival curve of 7-yr-old plantation was a skewed normal curve, and the survival rate and death rate were higher for I-age-class seedlings than others, and they were the highest for II-age-class seedlings. The survival curve of the 12-yr-old plantation showed Deevey-III type, that is, the early death rate was higher, the selection intensity of the environmental seize was greater, less than 10% of the seedlings could pass through this seize and enter the II-age-class. The transition from the seedling stage to the vegetative development stage was relatively balanced. Along with the increase of age, the environmental seize intensity at the vegetative development stage weakened, and the survival and death amount tended to level off (**Figure 5**).

2.5 Comprehensive evaluation of vegetation and soil system of *H. ammodendron* plantation

2.5.1 *The change characteristics of the vegetation and soil of* H. ammodendron *plantation*

The growth of *H. ammodendron* was rapid during 7 to 12 years after planting, and it began to decline after age 20, when the height, basal stem diameter, and crown width reached the maximum. During the age of 7 to 42, the tissue water content (TWC) and Proline mass fraction in the assimilated branches of *H. ammodendron* showed three curvilinear changes trend, i.e., increased from age 7 to 15, and then decrease slowly from age 30 to 35, after that increased again. Along with the increase of *H. ammodendron* age, the cell damage rate of *H. ammodendron* decreased inversely, the nitrogen mass fraction increased linearly and chlorophyll mass fraction increased exponentially. The critical change point of phosphorus mass fraction occurred at about the age of 25, showing a typical parabolic change characteristic. Meanwhile, the change trend of potassium mass fraction was irregular. For seven soil factors, only SHN (Y) had a power function regression relationship with plantation age (X) ($Y = 0.219X^{1.289}$, $F = 10.997$, $R^2 = 0.611$, $P = 0.013$), while SEP and SAP had no linear regression relationship with the plantation age. The EC increased along with plantation age, which might be caused by the accumulation and deposition of soluble salts in litters (**Table 5**).

2.5.2 *Classification system and evaluation criteria of vegetation and soil coupling type of* H. ammodendron *plantation*

According to the coupling coordination degree (D), combined with the comprehensive evaluation function of vegetation f(x) and soil environment g(y) calculated based on the classification method of Peng [29], the vegetation-soil coupling

Tree height class	7 yr			12 yr			17 yr			20 yr			28 yr			33 yr		
	a_x	q_x	e_x	a_x	q_x	e_x	a_x	q_x	e_x	a_x	q_x	e_x	a_x	q_x	e_x	a_x	q_x	e_x
I (0–0.5 m)	61	−180.3	2	375	960	0.7	943	814.4	1	236	830.5	0.9	2142	981.3	0.5	338	468.5	1.3
II (0.5-1 m)	72	777.8	0.8	15	133.3	5.6	175	57.1	2	40	625	1.9	40	975	1.2	180	758.1	1
III (1–1.5 m)	16	937.5	0.6	13	230.8	5.4	165	739.4	1.1	15	666.7	3.2	1	1000	26.5	43	702.9	1.5
IV (1.5-2 m)	1	0	1.5	10	−800	5.9	43	534.9	1.6	5	1000	7.7	0	0	0	13	−471.8	2.7
V (2–2.5 m)	1	0	0.5	18	166.7	2.5	20	200	1.9	0	0	0	0	0	0	19	789.5	1
VI (2.5-3 m)	/	/	/	15	200	1.9	16	562.5	1.3	1	−2000	35.5	1	−7000	25.5	4	250	2
VII (3–3.5 m)	/	/	/	12	250	1.3	7	428.6	1.2	3	−333.3	11.2	8	−250	2.6	3	0	1.5
VIII (3.5-4 m)	/	/	/	9	0	0.5	4	750	0.8	4	−3000	7.5	10	400	1.2	3	0	0.3
IX (4–4.5 m)	/		/	/	/	/	1	0	0.5	16	625	1.3	6	833.3	0.7	/	/	/
X (4.5-5 m)	/		/	/	/	/	/	/	/	6	500	1.5	1	0	0.5	/	/	/
XI (5–5.5 m)	/		/	/	/	/	/	/	/	3	0	1.5	/	/	/	/	/	/
XII (5.5-6 m)	/		/	/	/	/	/	/	/	3	0	0.5	/	/	/	/	/	/

Table 4.
Life table of six H. ammodendron *plantations at different chronosequence stage.*

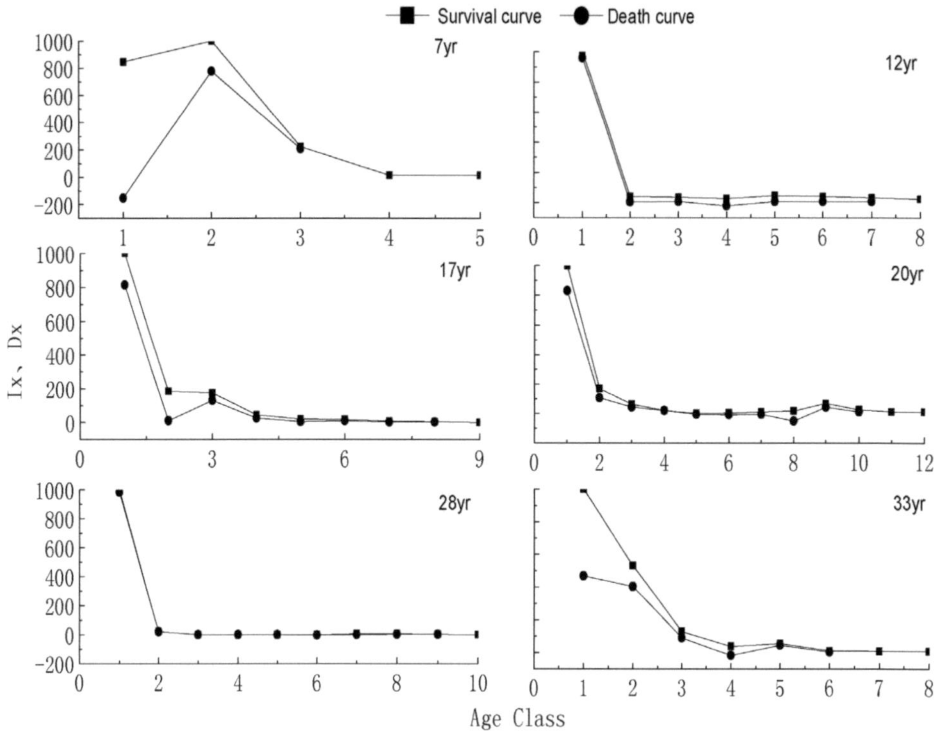

Figure 5.
Standard survival and death curves of H. ammodendron *plantations at different chronosequence stage.*

coordination type and evaluation criteria of *H. ammodendron* plantation were proposed, result shown in **Table 6**. If the ratio of f(x) to g(y) was bigger than 1, it indicated that the growth and development speed of vegetation was faster than that of soil. If the ratio was smaller than 1, it indicated that the growth and development speed of vegetation was slower than that of soil, and soil fertility resources were not fully utilized by the vegetation. If the ratio was closer to 1, it indicated that the succession state between them was more synchronous and coordinated. According to the D and the ratio between f(x) and g(y), the vegetation and soil coordinated development could be divided into 5 categories and 15 sub-categories. The coupling intensity and coupling coordination were not in one-to-one correspondence relationship, and both of them often appeared alternately, which was in line with the fluctuating development law between vegetation and soil subsystems.

2.5.3 Coupling coordination analysis of vegetation and soil system of H. ammodendron *plantation*

According to the data of **Table 7**, vegetation-soil coupling degree (C) of plantation was arranged on the order of 33 > 23 > 42 > 20 > 17 > 28 > 12 > 15 > 7. When *H. ammodendron* grew up to the age of 17, the interaction between vegetation and soil increased obviously and tended to be stable. The C of the 33-yr-old plantation was 73.85% increased than 7-yr-old one. The D of *H. ammodendron* plantations was arranged on the following order: 28 > 42 > 20 > 17 > 33 > 12 > 23 > 15 > 7. The D of 28-yr-old plantation was 96.63% increased than 7-yr-old one.

On the whole, the C and D of nine plantations showed a fluctuating and increasing trend along with the age increasing. *H. ammodendron* populations always kept at the level of disorder and declining, and vegetation and soil were in the state

Chronosequence stage(yr)	Vegetation comprehensive index f(x)											Soil comprehensive index g(y)						
	X1	**X2**	**X3**	**X4**	**X5**	**X6**	**X7**	**X8**	**X9**	**X10**	**X10**	**Y1**	**Y2**	**Y3**	**Y4**	**Y5**	**Y6**	**Y7**
7	0.60	0.89	0.12	0.05	43.3	0.28	49.41	0.53	1.25	0.17	2.14	0.93	0.84	2.97	1.80	58.93	8.38	0.09
12	2.81	5.69	4.37	3.82	63.71	0.26	52.56	0.58	1.59	0.09	1.26	6.15	3.06	4.02	2.04	173.5	9.31	1.62
15	2.84	6.08	5.93	4.88	65.66	0.25	40.04	0.94	1.22	0.11	1.5	5.49	0.91	6.72	1.86	64.4	8.03	0.16
17	2.64	7.22	3.05	5.57	61.61	0.36	43.86	0.61	1.43	0.087	2.69	5.14	4.08	15.9	3.68	163.6	7.93	0.76
20	4.29	10.67	11.09	16.17	64.3	0.23	36.60	0.82	1.79	0.11	2.75	1.52	2.74	10.79	3.99	254.8	8.86	0.89
23	2.15	5.59	3.50	3.57	61.67	0.27	27.55	1.06	1.34	0.09	1.27	3.05	1.22	6.88	1.83	84.3	9.67	0.6
28	3.68	9.59	9.24	11.00	62.12	0.23	36.60	0.82	1.79	0.11	2.74	4.33	2.91	31.52	6.85	275.5	8.49	2.09
33	2.37	5.48	3.04	3.59	60.73	0.23	36.60	0.82	1.79	0.11	2.74	3.36	3.94	7.46	3.14	152.6	7.88	0.33
42	2.55	8.70	6.41	8.52	68.75	0.50	33.23	1.06	1.83	0.18	1.92	1.48	1.46	45.32	2.74	124	8.84	1.24

X1: Height (m), X2: Basal stem diameter (cm), X3: Crown projected area of canopy storey ($m^2 \cdot tree^{-1}$), X4: Above-ground biomass of canopy storey ($kg \cdot tree^{-1}$), X5: Tissue water content (%), X6: Proline mass fraction ($ug \cdot g^{-1}$), X7: Cell damage rate(%), X8: Chlorophyll mass fraction ($mg \cdot g^{-1}$), X9: Nitrogen mass fraction(%), X10: Phosphorus mass fraction(%), X11: Potassium mass fraction(%). Y1: SWC(%), Y2: SOM ($g \cdot kg^{-1}$), Y3: SHN ($mg \cdot kg^{-1}$), Y4: SEP ($mg \cdot kg^{-1}$), Y5: SAP ($mg \cdot kg^{-1}$), Y6: pH value, Y7: EC ($ms \cdot cm^{-1}$).

Table 5.
The basal characteristic of vegetation and soil in H. ammodendron *plantations.*

Coupling coordination degree (D)	Type level	f(x)/g(y)	Coupling coordination type
0<D ≤ 0.1	Extreme imbalance of recession	>1.2	Highly disordered and declining class vegetation profit and loss type
		0.8 ~ 1.2	Highly disordered and declining class vegetation-soil co-loss type
		<0.8	Highly disordered and declining class soil profit and loss type
0.1<D ≤ 0.2	Serious imbalance of recession	>1.2	Seriously disordered and declining class vegetation profit and loss type
		0.8 ~ 1.2	Seriously disordered and declining class vegetation-soil co-loss type
		<0.8	Seriously disordered and declining class soil profit and loss type
0.2<D ≤ 0.3	Moderate imbalance of recession	>1.2	Moderately disordered and declining class vegetation profit and loss type
		0.8 ~ 1.2	Moderately disordered and declining class vegetation-soil co-loss type
		<0.8	Seriously disordered and declining class soil profit and loss type
0.3<D ≤ 0.4	Mild imbalance of recession	>1.2	Slightly disordered and declining class vegetation profit and loss type
		0.8 ~ 1.2	Slightly disordered and declining class vegetation-soil co-loss type
		<0.8	Slightly disordered and declining class soil profit and loss type
0.4<D ≤ 0.5	Brink imbalance of recession	>1.2	Nearly disordered and declining class vegetation profit and loss type
		0.8 ~ 1.2	Nearly disordered and declining class vegetation-soil co-loss type
		<0.8	Nearly disordered and declining class soil profit and loss type

Table 6.
Types of vegetation and soil coupling coordinated development.

of profit and loss. However, the degree of disorder on profit and loss showed the trend from serious to slight (**Table 7**).

2.5.4 Comprehensive evaluation of vegetation and soil system of H. ammodendron *plantation*

Through principal component analysis, the overall information of the original 18 variables were replaced by the first 5 principal components, contribution rate was 90.91% (**Table 8**). Among them, the Eigen Value of the first principal component was 7.531, which accounted for 41.84% of the total variation and formed a correlation with above-ground biomass (0.900), crown projected area (0.874), basal stem diameter (0.831), and tree height (0.830) etc.. Therefore this principal component could represent the growth characteristic index of vegetation. The Eigen Value of the second principal component was 3.486, which accounted for 19.37% of the total variables. The chlorophyll mass fraction (0.918) had a greater positive effect on it, and the cell damage rate (−0.858) had a strong negative correlation, so this

Chronosequence stage (yr)	f(x)	g(y)	C	D	f(x)/g(y)	Coupling coordination type
7	0.083	0.043	0.575	0.190	1.958	Seriously disordered and declining class vegetation profit and loss type
12	0.098	0.146	0.817	0.316	0.668	Slightly disordered and declining class soil profit and loss type
15	0.104	0.064	0.749	0.251	1.620	Moderately disordered and declining class vegetation profit and loss type
17	0.108	0.141	0.917	0.338	0.768	Slightly disordered and declining class soil profit and loss type
20	0.146	0.113	0.919	0.345	1.297	Slightly disordered and declining class vegetation profit and loss type
23	0.091	0.080	0.978	0.289	1.143	Moderately disordered and declining class vegetation-soil co-loss type
28	0.134	0.187	0.871	0.374	0.717	Slightly disordered and declining class soil profit and loss type
33	0.106	0.109	0.999	0.328	0.973	Slightly disordered and declining class vegetation - soil co-loss type
42	0.137	0.118	0.972	0.351	1.164	Slightly disordered and declining class vegetation - soil co-loss type

Table 7.
The comprehensive evaluation on coupling coordination of vegetation and soil system of H. ammodendron *plantations at different chronosequence stage.*

principal component mainly could reflect the physiological characteristics of the vegetation. The SMC (0.947) had the greatest positive effect on the third principal component, so the principal component could mainly reflect the physical characteristics of soil. The variable related to the fourth principal component was sHN (0.808), so this principal component could represented the nutrient characteristics of soil. The variable which had the greatest negative effect on the fifth principal component was potassium mass fraction of vegetation (−0.805), so this principal component could represent the nutrient characteristics of vegetation. The extracted five principal components could represent the variables including the growth, physiological and nutrient characteristics of vegetation, physical and chemical characteristics of soil. So these five principal components could be used to replace the comprehensive evaluation score of the identification and evaluation index system of *H. ammodendron* plantation.

Through the comprehensive evaluation scores of *H. ammodendron* plantation in **Table 9**, it could be seen that the 42-yr-old one ranked the first, followed by plantation with the age order of 28, 20, 23, 12, 15, 17, 33 and 7. That is to say, during the process of sand fixation, the comprehensive score calculated based on the five principal component scores increased gradually, thus the comprehensive ranking was gradually higher than before (**Table 9**).

2.6 Conclusions and discussion

The constructive species could regenerate successfully, which is the basic premise of sustainable development of artificial forest. The 7-yr-old *H. ammodendron* plantation, in the southern fringe of Gurbantunggut Desert, had weak self-generation ability and it could maintain such ability until the age of 33. During the process of individuals growth in canopy storey, there were abundant regenerated individuals of the grade I seedlings under canopy, which indicated that

Factor	Component				
	1	2	3	4	5
Tissue water content (%)	0.389	0.651	0.49	0.275	−0.169
Proline mass fraction (ug·g^{-1})	−0.155	0.167	−0.111	0.907	−0.027
Cell damage rate (%)	−0.184	−0.858	0.162	−0.067	−0.005
Chlorophyll mass fraction (mg·g^{-1})	0.074	0.918	−0.142	0.188	−0.201
Nitrogen mass fraction (%)	0.766	0.093	0.044	0.345	0.156
Phosphorus mass fraction (%)	−0.089	−0.147	−0.821	0.495	0.068
Potassium mass fraction (%)	0.533	−0.128	−0.143	−0.019	−0.805
Height(m)	0.830	0.356	0.319	−0.152	−0.022
Basal stem diameter (cm)	0.831	0.457	0.198	0.155	0.017
Crown projected area (m^2·tree^{-1})	0.874	0.378	−0.034	−0.082	−0.127
Above-ground biomass (kg·tree^{-1})	0.900	0.325	−0.094	−0.03	0.02
SWC(%)	0.071	−0.238	0.947	−0.02	−0.076
SHN (mg·kg^{-1})	0.469	0.247	−0.14	0.808	0.036
SEP (mg·kg^{-1})	0.83	−0.64	0.084	0.056	0.331
SAP (mg·kg^{-1})	0.949	−0.123	0.231	−0.068	0.129
SOM(g·kg^{-1})	0.402	−0.216	0.647	0.001	0.515
EC (ms·cm^{-1})	0.742	−0.203	0.325	0.352	−0.348
pH value	0.121	0.087	−0.007	−0.17	−0.898
Eeigen value	7.531	3.486	2.469	1.619	1.259
Relative variance contribution (%)	41.840	19.369	13.715	8.996	6.994
Accumulated Variance Contribution Rate (%)	41.840	61.208	74.923	83.918	90.912

Table 8.
The Eigen values and contribution rate of the principal constituent.

Chronosequence stage (yr)	F1	F2	F3	F4	F5	Comprehensive Evaluation score	Ranking
7	−1.869	−0.133	1.422	−0.135	1.209	0.210	9
12	−0.160	−1.092	−1.365	1.401	0.935	0.723	5
15	−0.563	0.430	−0.724	−0.883	−0.701	0.601	6
17	0.051	−1.046	0.324	0.679	−1.415	0.471	7
20	1.239	0.009	0.181	−1.465	0.923	0.815	3
23	−0.549	0.924	−1.561	−0.541	−0.009	0.739	4
28	1.440	−0.474	0.449	0.148	0.662	0.897	2
33	−0.085	−0.627	0.589	−0.602	−1.345	0.425	8
42	0.496	2.010	0.683	1.398	−0.261	0.918	1

Table 9.
Comprehensive evaluation of H. ammodendron *plantations.*

the *H. ammodendron* seeds can successfully germinated and grew into seedlings, as well as reflected the characteristics of high maturing rate, short seed dormancy period and high germination rate of the *H. ammodendron*. Which was also the main reason for abundance of grade I seedling (HT < 30 cm). The number of grade I seedlings accounted for the largest proportion in regeneration storey, and rich seedling bank could provide a strong guarantee for the regeneration and succession of the population [30]. However, the large seedling bank could not guarantee proper regeneration and ensure sufficient number of higher grade seedlings, and grade III seedlings (HT ≥ 50 cm) were the key for transferring from the regeneration storey to the canopy storey, and thus realizing natural regeneration and sustainable development of the *H. ammodendron* population. Therefore, moderate thinning and tending measures should be taken for *H. ammodendron* plantation to form a more stable structure. The SHN was the main influencing factor of seedling regeneration in *H. ammodendron* plantation. Therefore, avoiding destroy the shrub and grass with strong nitrogen fixation ability, and properly retaining withered litters on soil surface would increase the SHN content in soil, and at last would promote the growth of seedlings.

The *H. ammodendron* plantation system, at the south edge of Gurbantunggut Desert, had weak self-optimization ability, the interdependence and mutual restriction relationship between vegetation and soil. Along with the increase of the plantation age, the ratio of f(x)/g(y) gradually increased too. It showed that *H. ammodendron* population restoration measures had a strong ability to adapt to the environment, and the growth and development speed of population exceeded the development speed of soil, and this may indicate that such vegetation was able to adapt to the local environment, but had a limited ability to reconstruct the soil. When the plantation developed to the age of 42, the coupling degree (C) and coupling coordination degree (D) between vegetation and soil system, f(x)/g(y) were relatively larger, it could be inferred that the vegetation-soil system of the *H. ammodendron* plantation tended to transform from a declining type to a coordinated development type. It is concluded that the competition of vegetation growth would be more intense in the coming years. Therefore, artificial management measures, such as spring water collection, regular fertilization and tree pruning are needed to promote the healthy growth of *H. ammodendron* and the natural regeneration of seedlings, so as to improve the diversity and coverage of vegetation, in order to enhance the coordination degree of vegetation and soil, and finally realize the transformation to a high-level coupling state.

Author details

Qinghong Luo[1*], Qimin Chen[1], Miao He[1] and Na Li[2]

1 Institute of Afforestation and Sand Control, Xinjiang Academy of Forestry Sciences, Urumqi, China

2 Xinjiang Forestry Ecological Monitoring Station, Urumqi, China

*Address all correspondence to: lqh482325@sina.com

References

[1] Johnson, J.B. 1997. Stand structure and vegetation dynamics of a subalpine wooded fen in Rocky Mountain National Park, Colorado. J. Veg. Sci., 8 (3): 337-342.

[2] Agren, J., and Zackarisson, O., 1990. Age and size structure of *Pinus sylvestris* populations on mires in Central and Northern Sweden. J. Ecol., 78: 1049-1062.

[3] Song, C.W., Liu C.J., Halik Ü., et al. 2021. Spatial Distribution and Structural Characteristics for Haloxylon ammodendron Plantation on the Southwestern Edge of the Gurbantünggüt Desert. Forests, 12(5), 633-633.

[4] Liu, G.J., Zhang, X.M., Li, J.G., Wu, J.X., Wei, J., Li, J.L., Wang, X.W. 2010. Analysis on dynamic life table of natural Haloxylon ammodendron at seedling stage in the southeastern marginal zone of the Junggar Basin. Arid Zone Res., 27 (1): 83-87.

[5] Qian, Y.B., and Li, Y.F. 1999. Study on soil physical properties of desertified land in Mosuowan Reclamation Area. Arid Zone Res., 26(2): 31-37.

[6] Zhou, C.B., Song, Y.Y. 2015. Influence of different longitudinal dune positions in the Gurbantunggut Desert on the reproduction of Haloxylon ammodendron seedlings. Appl. Ecol. Env. Res., 13(1): 99-113.

[7] Magnus-Lof, D.C., Dey-Rafael, M.N., Douglass, F.J. 2012. Mechanical site preparation for forest restoration. New. For., 43: 825-848.

[8] Vacek, S.I., Nosková, L., Bílek, Z., Vacek, O.S. 2010. Regeneration of forest stands on permanent research plots in the Krkonoše Mts. J. For. Sci., 56(11): 541–554.

[9] Zhang, X.Y., Wang, X.M., Yan, P. 2007. Re-evaluating the impacts of human activity and environmental change on desertification in the Minqin Oasis, China. Env. Geo., 55(4): 705-715.

[10] Wang G.H., Zhao W.Z. 2015. Effects of seed density on the germination and seedling growth of Haloxylon ammodendron. Journal of Desert Research, 35(5): 1248-1253.

[11] Vander K.J., Yassir I., Buurman P. 2009. Soil carbon changes upon secondary succession in Imperata grasslands (East Kalimantan, Indonesia) [J]. Geoderma. 149(1-2): 76-83.

[12] SU Y.Z., and LIU T.N. 2020. Soil Evolution Processes Following Establishment of Artificial Sandy-fixing Haloxylon Ammodendron Forest. Acta Pedologica Sinica, 57(1): 84-91.

[13] Karlen D.L., Rosek M.J., Gardner J. C., et al. 1999. Conservation reserve program effects on soil quality indicators[J]. Journal of Soil and Water Conservation, 54: 439-444.

[14] Research group of "study on combating desertification/degradation in china. 1998. Study on combating desertification/degradation in China. Beijing: Environmental Science Press.

[15] Zhou, Z.B., Xu, X.W. 2002. Absorption characteristics of ions in three species of shrubs of the artificially-planted greenbelts in the Hinterland of Taklamakan Desert. Arid Zone Res., 19 (1): 49-52.

[16] Han, W., Xu, X.W., Li, L., Lei, J.Q., Li, S.Y. 2010. Chlorophyll a fluorescence responses of *Haloxylon ammodendron* seedlings subjected to progressive saline stress in the Tarim Desert highway ecological shelterbelt. Photosynthetica, 48(4): 635-640.

[17] Arabzadeh, N. 2011. A Study on the impact of drought stress on changes of free proline in two species of Haloxylon persicum and Haloxylon aphyllum. J. Plant Sci., 6(5): 190-201.

[18] Li, Y.F. 1992. Effects of different moisture habitats on water consumption of Haloxylon ammodendron. Arid Zone Res., 9(4): 45-50.

[19] Wang, X.Q., Wang, T., Jiang, J., Zhao, C.J. 2005. On the sand surface stability in the southern part of Gurbantunggut Desert. Science in China Series D: Earth Sciences, 48(6): 778-785.

[20] Huang, P.Z., Liu, Z.J., Cui, W.C. 1985. A preliminary study on water harvesting afforestation of Haloxylon ammodendron. Xinjiang Agricultural Sciences, 23-25.

[21] Qian, Y.B., and Li, Y.F. 1999. Study on soil physical properties of desertified land in Mosuowan Reclamation Area. Arid Zone Res., 26(2): 31-37.

[22] Phillips, D.L., and MacMahon, J.A. 1981. Competition and spacing patterns in desert shrubs. J. Ecol., 69: 97-115.

[23] Song, Y.Y., and Hu, X.J. 2011. Estimation models of the aboveground biomass of *Haloxylon ammodendron* in the different ecotype of the Gurbantunggut Desert. J. Northwest For. Univer., 26(2): 31-37.

[24] Rundel, P. 1971. Community structure and stability in the giant sequoia groves of the Sierra Nevada. Am. Midl. Nat., 85:478-492.

[25] Liu, G.J., Zhang, X.M., Li, J.G., Wu, J.X., Wei, J., Li, J.L., and Wang, X.W. 2010. Analysis on dynamic life table of natural *Haloxylon ammodendron* at seedling stage in the southeastern marginal zone of the Junggar Basin. Arid Zone Res., 27(1): 83-87.

[26] Blackman, G.E. 1942. Statistical and ecological studies in the distribution of species in plant communities: I. Dispersion as a Factor in the Study of Changes in Plant Populations. Ann. Bot., 6: 351-370.

[27] China's State Forestry Administration (SFA). 2010. Forest resources planning survey, GB/T 26424 State Forestry.

[28] Li, F.J., Sun, D.X., Chang, Z.F. 2008. Preliminary study on natural regeneration mechanism of *H. ammodendron* forest in Minqin desert area. Chinese Agricultural Science Bulletin, 24(9): 165-170.

[29] Peng, W.X., Song T.Q., Zeng, F.P. et al. 2011. Model of vegetation and soil coupling coordinative degree in grain for green project in depression between karst hills. Transactions of the CSAE, 27 (9): 305-310.

[30] Li G.L., Zhang D.H., Zhang Z.S., et al. 2020. Population dynamics of main sand-fixing shrubs in the Gurbantunggut Desert. J. of Desert Research, 41(2):129-137.

Chapter 7

Evaluation of Soil Seed Banks in Different Aged *Caragana microphylla* Plantation in Desert Steppe Ecosystems

Guoqi Li, Shujun Li, Wenshan Shao, Yanyun Chen and Wang Yafang

Abstract

Soil seed bank (SSB) represents potential plant communities, which is essential in the restoration of degraded ecosystems. Consequently, SSB is crucial in the reconstruction and recovery of aboveground plants because they largely determine the process and direction of vegetation restoration. SSB is also important indicators that can be used to evaluate the effects of management on degraded desert steppe. Here, field sampling and soil seed germination experiments were used to investigate the role of SSB in the recovery of degraded desert steppe. Results indicated that (1) the species composition of SSB and ground vegetation significantly differed in different aged *Caragana microphylla* plantation and control in the Yanchi County. (2) The abundance of SSB was significantly promoted by *C. microphylla* plantation. The average seed density in *Caragana* plantation SSB was 11248.75 m^{-2}, which was 17 times than that of SSB in areas without *C. microphylla* plantation. (3) The ages of *C. microphylla* plantation were closely related to the composition and density of SSB.

Keywords: desert steppe, soil seed bank (SSB), species composition, vertical distribution pattern, species diversity

1. Introduction

The soil seed bank (SSB) comprises seeds that have survived at the soil surface and in the soil [1]. The SSB represents potential plant communities and is essential for rehabilitating degraded ecosystems via revegetation and restoration. Moreover, the SSB determines the progress and direction of the future vegetative composition of communities [2–4]. Consequently, understanding species composition, density, and diversity of SSB can provide a basis for understanding mechanisms of vegetation succession, and is a particularly important indicator for evaluating the effects from treating degraded desert steppe ecosystems [5]. Human disturbance always generate significant negative effects on the SSB in desertified regions in a semi-arid climate [6]. The research on seed banks in a degraded desert shrubland showed there were consistently positive relationships between ground cover of litter and viable seed density [7]. In degraded drylands, the re-establishment of post-disturbance native plant community was almost exclusively conducted using seeds [8]. The average seed density decreased from stabilized sand dune to interdune lowland of stabilized sand dune, to interdune lowland of active sand dune, and to active sand dune [9]. The latest researches indicated that vegetation cover and seed bank size were notably lower in the degraded grassland. And there were significant differences in understory vegetation and soil seed stocks among different vegetation types on the Loess Plateau [10]. The restoration effect of planted forest was obviously better than that of abandoned farmland [11]. Tree seed density and diversity increased with regeneration time and vegetation structure [12]. Spatially, the distribution of the seed density and species diversity of the SBB in the marginal zones were higher than those of seed banks in the hinterland of the desert [13, 14]. Although considerable research has investigated various aspects of SSB characteristics for different vegetation types, little research has been conducted to systematically evaluate SSB of planted shrubs, and how those treatments can affect dessert steppe ecosystems.

Beginning in the 1970s, plantation of *C. microphylla* in addition to fencing has been used in Yanchi County to restore degraded desert steppe. Currently, there are 600,000 hectares *C. microphylla* plantation. In this study, the SSB was investigated in semi-fixed dune, and *C. microphylla* plantation with different ages within fencing in a degraded desert steppe, while using a neighboring natural steppe without desertification as a control. Several important indicators of SSB were evaluated, including species composition, the number of germinations, and species diversity. The changes of plant species and seed density in SBB of *C. microphylla* plantation with different ages can provide scientific basis for the selection of plant species and the determination of seeding amount for aerial seeding [15].

2. Material and methods

2.1 Study site and plot layout

The study area was based in the Ningxia Hui Autonomous Region of Eastern China (37°04′ $\sim$ 38°10′N, 106°30′ $\sim$ 107°41′E) and features an elevation of 1,295 $\sim$ 1,951 m. Yanchi County is connected to the Mu Us Sandland in the north and to the Loess Plateau in the south, which is considered to be a typical geographic transition zone. Desertification areas are expanding in Yanchi County due to overuse of ecosystem resources. Sand land has become one of the main landform types in the Ordos slope hills of the area. The total area of Yanchi County is 8,661.3 km^2. It has a typical temperate continental climate with hot summer, cold winter, extensive drought, significant wind, and strong evaporation. The average annual temperature was 8.1°C.The average annual rainfall was 250–350 mm, wherein greater than 80% of the rainfall occurred from May to September. The primary winds in the area were west and northwest winds. The main soil of Yanchi County is sierozem, followed by black loam and eolian soil [16]. The primary vegetation in Yanchi County is classified as a transition zone between the central Asian sub-regions of Eurasian steppe zone and the central steppe region. Vegetation types in the area include shrubs, steppe, meadows, swamps, and deserts. Vegetation exhibits transitional characteristics with gradual succession that is staggered extending from the south to the north. Yanchi County contains only plantation forests that mainly consist of *Salix psammophila* shrubland and *C. microphylla* shrubland [17]. The survey included semi-fixed dunes without *C. microphylla* plantation (plot ID: NO.1), and soil substrates were sierozem [18]. The dominant species in Plot 1 are *Agriophyllum squarrosum* and *Setaria viridis*. *C. microphylla* plantation with 7, 16, 25, and 37 years were included in the study and are identified as plots NO.2–5, respectively. Lastly, a neighboring natural steppe that did not exhibit desertification was used as a control (NO.6) (**Figure 1**). The dominant species in Plot 6 are *Artemisia scoparia* and *Sophora alopecuroides.*

2.2 Ground vegetation survey

Typical sampling methods were used in mid-August of 2015 to survey above-ground plant species composition, densities, heights, coverage, and frequencies in the various *C. microphylla* plantations. The specific location of plants in the plots, micro-topography, and other characteristics were simultaneously recorded in addition to sub-cover via grid needling methods. Each of the six plots consisted of ten 1 m × 1 m quadrats for a total of 60 quadrats.

2.3 Soil seed bank analyses

Transect line sampling was conducted, as previously described [19, 20]. Specifically, random arrangement of two 10 m long sample lines were placed in each sample plot in April 2013 using a customized soil seed bank sampler with equidistant 1 m sections that were then sampled at 0–2 cm, 2–5 cm, and 5–10 cm depths. Soil samples (n = 360) were placed in bags and transported back to the laboratory where they were then used for soil seed bank germination experiments. Indoor soil seed germination experiments were conducted as described previously [21]. The sampling area of each sample plot was 0.8 m^2, and the total sampling area of the six sample plots was 4.8 m^2. Briefly, direct germination measurements were used to estimate the portion of the seed bank capable of producing germinating seeds and the species composition of the SSB. Seedling germination can usually detect more than 90% of the species within the SSB of grassland ecosystems [22]. Soils were

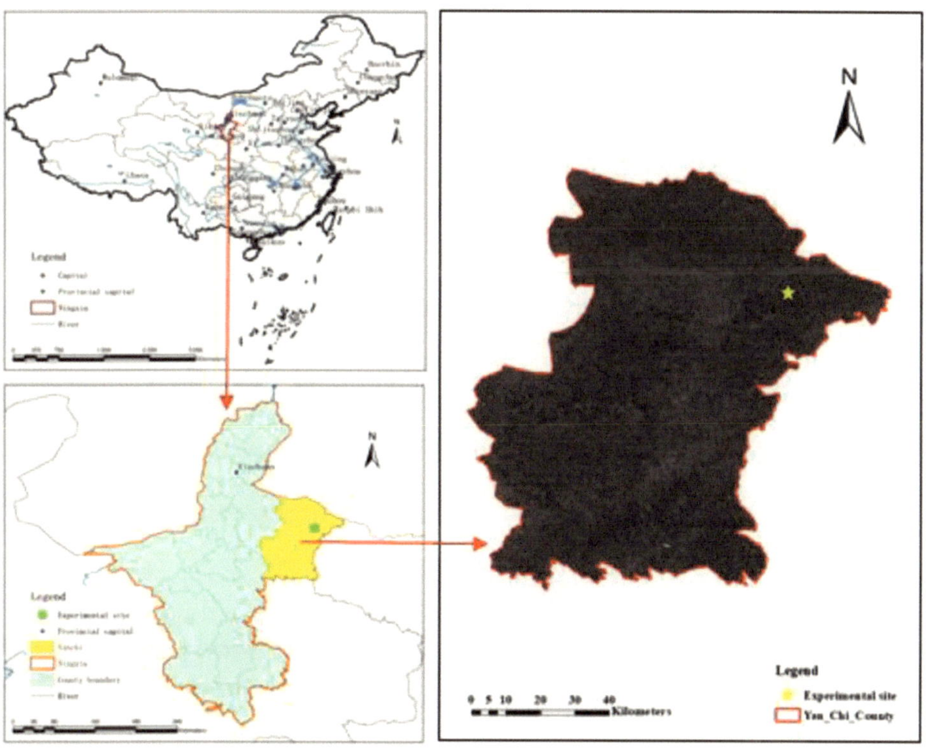

Figure 1.
The locations of the sampling sites.

sun-dried inside of a greenhouse then sieved to remove litter, roots, rocks, and other debris. Soil samples were evenly spread and then placed in pots (42 cm × 28 cm × 15 cm) with a soil thickness of 1.5 cm with seed-less sand used as matrix at the bottom of the pots at a height of 10 cm.

2.4 Statistical analyses

Seed density was measured as the number of germinated plants in the soil seed bank of each plot, which was then converted to a seed bank number (in 1 m^2 units). Species diversity of the SSB was calculated based on seed densities and species composition via the Shannon-Wiener diversity index (H), the Simpson diversity index (D), and the Pielou evenness index (E) [23–25], as follows

$$\text{Richness index}: \mathrm{R} = (\mathrm{S}-1)/\ln\mathrm{N} \tag{1}$$

$$\text{Shannon} - \text{Wiener diversity index (H)}: H = -\sum_{i-1}^{s}(P_i \, 1\mathrm{n} P_i) \tag{2}$$

$$\text{Simpson diversity index}: D = 1 - \sum_{i=1}^{S} P_i^2 \tag{3}$$

$$\text{Pielou evenness index (E)}: E = H/1\mathrm{n}S \tag{4}$$

Where N is the total number of seeds in the SSB for each plot type, S is the total number of species in the SSB of each plot, and Pi is the number of the i-th species' seeds divided by the total number of seeds in a given plot.

The SPSS19.0 software package was used for single-factor analysis of variance (one-way ANOVA) tests of SSB differences, while the least significant difference method (LSD) was used to evaluate the statistical significance of differences. Microsoft Excel 2003 was used for graphical representations and other statistical analyses.

3. Results

3.1 Community characteristics of different age *C. microphylla* plantation

Ground vegetation communities within different aged *C. microphylla* plantation belonged to 10 families and 31 genera (**Table 1**), in which the Asteraceae (seven species), and Leguminosae (seven species) comprised the majority of plants, followed

Plot	Species	Coverage	Dominant species	Companion species
NO.1	5	< 1%	*Agriophyllum squarrosum, Setaria viridis*	*Agropyron desertorum*
NO.2	18	39.80%	*Artemisia scoparia, Corispermum mongolicum*	*Cynanchum komarouii, Salsola beticolor Setaria viridis*
NO.3	31	61.90%	*A. scoparia*	*A. desertorum, Lespedeza potaninii*
NO.4	17	71.70%	*A. scoparia*	*C. microphylla, Pennisetum centrasiaticum*
NO.5	16	49.80%	*A. scoparia*	*Setaria viridis, Sophora alopecuroides*
NO.6	12	69.50%	*A. scoparia, Sophora alopecuroides*	*Lespedeza potaninii, Euphorbia esula*

Table 1.
Community characteristics among 6 plots of C. microphylla *plantation and control.*

by the Gramineae (six species), the Chenopodiaceae (four species), and Convolvulacese (three species) in addition to Euphorbiaceae, Asclepiadaceae, and Zygophyllaceae that each comprised two species, and the Liliaceae and Polygalaceae that each constituted only one species. The species were divided into three kinds of life forms, including 3 shrubs or semi-shrubs (*Sophora alopecuroides*, *Caragana microphylla*, and *Lespedeza potaninii*), 11 annual plants, and 21 perennial herbaceous plants.

3.2 Species composition and seed density of SSB in different aged *C. microphylla* plantation

There were 33 species of plants in the soil seed bank and 35 species of above-ground vegetation in all sample plots in the study area, among which 18 species of plants were found in both SBB and above-ground vegetation. Among the plots, 15 species only appeared in SSBs, while 17 species only appeared in the aboveground vegetation communities. A total of 33 plant species were identified in the study area that belonged to 13 families and 27 genera. The highest number of species were present in Gramineae (10) followed by Chenopodiaceae (7), and then Leguminosae (3) in addition to two species each for Asteraceae, Euphorbiaceae and Asclepiadaceae, and then one species each for Plantaginaceae, Zygophyllaceae, Rosaceae, Caryophyllaceae, Amaranthaceae, Convolvulacese, and Polygalaceae. In terms of plant life form, two species of semi-shrubs were identified (*Sophora alopecuroides* and *Lespedeza potaninii*), in addition to 11 perennial herbs and 20 species of one or two years herbaceous plants (**Table 2**).

In the semi-fixed sand dunes, there were 661.25 / m^2 germinated seeds of the SSB consisting of 13 plant species that were dominated by *Setaria viridis* and *Digitaria ischaemum*, which accounting for 70.13% and 13.42% of plant community composition, respectively. In the 7 years *C. microphylla* plantation, there were 12,290.00·m^{-2} seeds comprising 21 plant species that were dominated by *Artemisia scoparia*. The SSB seed number decreased to 10,553.75·m^{-2} in the 16 years *C. microphylla* plantation, but there were 29 species that were mostly represented by nine species of Gramineae that accounted for 12.59% of the total plant composition. The SSB density of the 25 years *C. microphylla* plantation was 15,891.25·m^{-2}, which was highest among the SSB of *C. microphylla* plantation, although low species evenness was observed, with *Artemisia scoparia* accounting for 93.81% of the plants in these plots. The SSB density in the 37 years *C. microphylla* plantation was 6,260.00·m^{-2}, which was significantly lower than in other *C. microphylla* plantation. This decreased density in the oldest forest may be related to the types of species present in these plots and significantly reduced ground vegetation coverage (**Table 2**).

The average density of germinated seeds from SSBs in *C. microphylla* plantation was 11,248.75, with a total range from 6,262.00 - 15,891.25·m^{-2}. SSB seed numbers increased with forest age rapidly, followed by slight decreases with age after that. The mid-aged forests thus had the highest seed densities, while the density of the SSB in semi-fixed sand dunes was 661.25·m^{-2}. The difference equates to a 17-fold increase in seed density in planted area SSBs compared to non-planted areas (**Table 2**).

The densities of the SSB in *C. microphylla* plantation were lower than that of the natural grassland (19,788.75·m^{-2}). However, there were less species in the SSB of the natural grassland (22), relative to the 16 and 37 years *C. microphylla* plantation (**Table 2**).

3.3 Vertical distributions of species numbers of SSB in different aged *C. microphylla* plantation

The vertical distribution of seeds within the SSB is an essential component for SSBs within degraded desert steppe ecosystems. With increasing age of

Species	Plot1		Plot 2		Plot 3		Plot 4		Plot 5		CK	
	Composition/ %	Density/ Seeds·m^{-2}	Composition/ %	Density/ Seeds·m^{-2}	Composition/ %	Density/ Seeds·m^{-2}	Composition/ %	Density/ Seeds·m^{-2}	Composition/ %	Density/ Seeds·m^{-2}	Composition/ %	Density/ Seeds·m^{-2}
Plantago asiatica	—	—	—	—	—	—	0.01	1.25	—	—	—	—
Euphorbia humifusa	0.19	1.25	0.19	23.75	0.06	6.25	0.03	5.00	0.06	3.75	0.04	8.75
Euphorbia esula	—	—	0.02	2.50	0.01	1.25	—	—	—	—	0.01	1.25
Sophora alopecuroides	—	—	0.02	2.50	0.04	3.75	0.02	3.75	—	—	0.01	1.25
Gueldenstaedtia stenophylla	—	—	—	—	0.18	18.75	0.04	6.25	0.02	1.25	—	—
Lespedeza potaninii	—	—	0.06	7.50	2.55	268.75	0.06	8.75	0.12	7.50	0.02	3.75
Tragus racemosus	—	—	—	—	0.75	78.75	0.96	152.50	2.94	183.75	—	—
Setaria viridis	70.13	463.75	1.13	138.75	2.27	240.00	0.96	152.50	1.00	62.50	0.25	50.00
Enneapogon borealis	—	—	—	—	1.40	147.50	—	—	0.08	5.00	—	—
Chloris virgata	0.19	1.25	0.01	1.25	0.01	1.25	—	—	0.06	3.75	0.01	1.25
Eragrostis poaeoides	0.76	5.00	3.25	400.00	4.18	441.25	1.09	173.75	5.03	315.00	0.20	38.75
Setaria glauca	—	—	0.70	86.25	1.07	112.50	0.08	12.50	1.14	71.25	—	—
Eleusine indica	4.35	28.75	0.27	33.75	0.01	1.25	0.03	5.00	0.06	3.75	0.11	22.50
Stipa glareosa	—	—	—	—	0.28	30.00	0.03	5.00	0.02	1.25	0.01	1.25
Digitaria ischaemum	13.42	88.75	1.45	178.75	2.62	276.25	0.57	90.00	2.40	150.00	0.86	170.00
Tribulus terrestris	—	—	0.05	6.25	0.08	8.75	0.02	2.50	—	—	0.03	5.00
var. *graminifolia*	—	—	0.07	8.75	—	—	0.01	1.25	0.08	5.00	0.01	1.25
Artemisia scoparia	3.40	22.50	90.78	11156.25	82.27	8682.50	93.81	14907.50	83.23	5210.00	97.85	19362.50
Salsola ruthenica	0.57	3.75	0.05	6.25	0.02	2.50	0.03	5.00	0.08	5.00	0.01	1.25

Species	Plot1		Plot 2		Plot 3		Plot 4		Plot 5		CK	
	Composition/ %	Density/ Seeds·m^{-2}	Composition/ %	Density/ Seeds·m^{-2}	Composition/ %	Density/ Seeds·m^{-2}	Composition/ %	Density/ Seeds·m^{-2}	Composition/ %	Density/ Seeds·m^{-2}	Composition/ %	Density/ Seeds·m^{-2}
Chenopodium glaucum	5.67	37.50	0.39	47.50	0.32	33.75	0.37	58.75	0.98	61.25	0.35	70.00
Corispermum tylocarpum	0.19	1.25	—	—	—	—	—	—	—	—	—	—
Corispermum mongolicum	0.38	2.50	0.07	8.75	—	—	—	—	0.02	1.25	0.01	2.50
Agriophyllum squarrosum	—	—	0.58	71.25	0.04	3.75	—	—	—	—	0.05	10.00
Bassia dasyphylla	—	—	—	—	—	—	—	—	—	—	0.03	5.00
Salsola beticolor	0.38	2.50	0.72	88.75	1.02	107.50	1.86	295.00	1.32	82.50	0.08	15.00
Cynanchum chinense	—	—	—	—	—	—	—	—	—	—	0.01	1.25
Cynanchum komarouii	—	—	0.02	2.50	—	—	—	—	0.02	1.25	—	—
Potentilla chinensis	—	—	—	—	0.07	7.50	—	—	0.02	1.25	—	—
Stellaria media	—	—	—	—	0.33	35.00	—	—	—	—	—	—
Amaranthus retroflexus	—	—	0.05	6.25	0.02	2.50	0.01	1.25	0.04	2.50	0.08	15.00
Cuscuta chinensis	0.38	2.50	0.09	11.25	0.25	26.25	0.02	3.75	1.22	76.25	0.01	1.25
Polygala tenuifolia	—	—	—	—	0.15	16.25	—	—	—	—	—	—
Unknown	—	—	—	—	—	—	—	—	0.08	5.00	—	—
Total	100.00	661.25	100.00	12290.00	100.00	10553.75	100.00	15891.25	100.00	6260.00	100.00	19788.75

Table 2.
Average species compositions and seed densities in the SSB of different aged C. microphylla *plantation and control.*

C. microphylla plantation, the overall richness of species in the SSB increased dramatically, then decreased slightly (**Figure 2**). Within 0–2 cm soils along the age gradient of semi-fixed dunes to older *C. microphylla* plantation, the species richness was 10, 18, 21, 20 and 20, respectively. Species richness trended similarly along the age gradients for soils within the 2–5 cm, 5–10 cm, and 0–10 cm. The level of species richness was higher in the *C. microphylla* plantation compared to the natural steppe for all soil depths, with the exception of those of the 25 years *C. microphylla* plantation. The number of species in the 2–10 cm soil layer was 3, 3, 4, 0, 3, and 4, respectively.

3.4 Vertical distribution of seed densities of SSB in the different aged *C. microphylla* plantation

The total level of germinated seeds in the SSB of different aged *C. microphylla* plantation followed the progression of: 25 years >7 years >16 years >37 years. The number of germinated seeds significantly differed among all plots, except for between the 7 and 16 years. As was observed for SSB species richness, the number of SSB germinating plants first increased in younger plantation, then declined. Seed densities were primarily concentrated in the 0–2 cm soils, followed by the 2–5 cm depth soils and then the 5–10 cm depth soils. The number of seeds in the 0–2 cm SSB of the semi-fixed dune was 227.50 seeds·m^{-2}, which accounted for 34.40% of the 0–10 cm seed abundance, and was significantly lower ($P < 0.05$) than in other plots. Thus, these analyses focused on the seeds below 2 cm in the semi-fixed dune plots, since there were considerably lower seed numbers in the 0–2 cm depth interval. The density of seeds in the 0–2 cm depths of the 7, 16, 25, and 37 years *C. microphylla* plantation were 9,665.00 seeds·m^{-2}, 8,216.25 seeds·m^{-2}, 12,981.25 seeds·m^{-2} and 4,986.25 seeds·m^{-2}, respectively, which accounted for 78.64%, 77.85%, 81.69%, and 79.65% of the total 0–10 cm SSB. Significant differences were observed for the seed densities within the 0–2 cm soils of each forest plot. These results thus indicate that planting *Caragana* had a positive enrichment effect on the seed density in the 0–2 cm surface layer of SSBs, and that the strength of this effect varied with *C. microphylla* plantation age wherein: seed density in the 25 years >16 years >7 years >37 years. Compared to the natural grassland plots, both the germination total seeds and densities in each vertical soil interval were significantly lower in the planted forest plots, and the degree of difference varied among plots (**Table 3**).

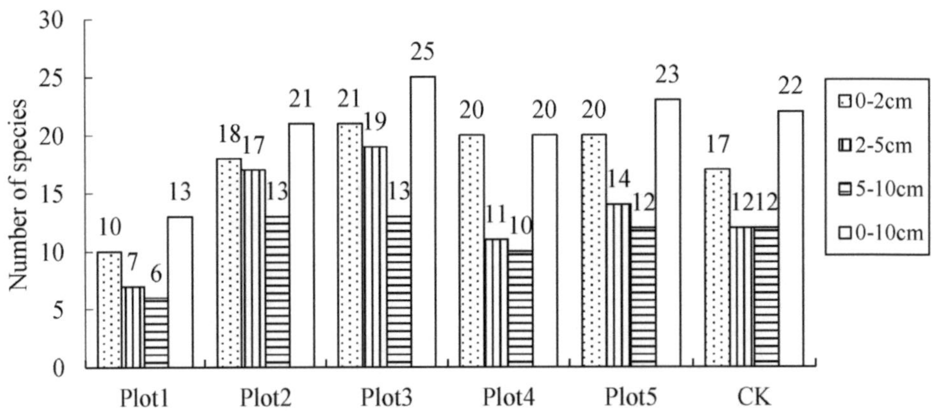

Figure 2.
Vertical distribution of species in the SSB of different aged C. microphylla *plantation and control.*

Sample Number	Germinated seeds of SSB(seeds·m^{-2})			Total seeds of SSB (seeds·m^{-2})
	0-2 cm	2-5 cm	5-10 cm	
NO.1	227.50 ± 84.85 d	246.25 ± 15.91 d	187.50 ± 127.28 b	661.25 ± 196.22 e
NO.2	9665.00 ± 1834.94 b	1651.25 ± 302.29 b	973.75 ± 47.73 ab	12290.00 ± 2184.96 c
NO.3	8216.25 ± 1568.01 b	1395.00 ± 144.96 bc	942.50 ± 102.53 ab	10553.75 ± 1815.50 c
NO.4	12981.25 ± 97.23 a	1556.25 ± 362.39 b	1353.75 ± 680.59 ab	15891.25 ± 220.97 b
NO.5	4986.25 ± 15.91 c	931.25 ± 199.76 c	342.50 ± 148.49 b	6260.00 ± 332.34 d
NO.6	15203.75 ± 924.54 a	2407.50 ± 91.92 a	2177.50 ± 1074.80 a	19788.75 ± 242.18 a

Note: The different letters indicates significant difference among plot at 0.05 level.

Table 3.
Soil seed bank vertical distribution in different aged C. microphylla *plantation and control.*

3.5 Species diversity of SSB in different aged *C. microphylla* plantation

The species richness of SSBs in *C. microphylla* plantation first increased with age, but then decreased (**Figure 3**). The number of species was highest in the 16 years *C. microphylla* plantation (22), and species richness was significantly different between each older plantation and the semi-fixed sand dunes. However, there was no significant difference between older plantation and the natural steppe ($P > 0.05$).

The Shannon-Wiener and Simpson indices trended similarly, with each value increasing as plantation age increased, but with different degrees of difference among each plantation age. The Pielou evenness index varied as plantation age increased. The number of species in the natural grassland (16.500 ± 3.536) was lower than the number of species in each of *C. microphylla* plantation. Meanwhile,

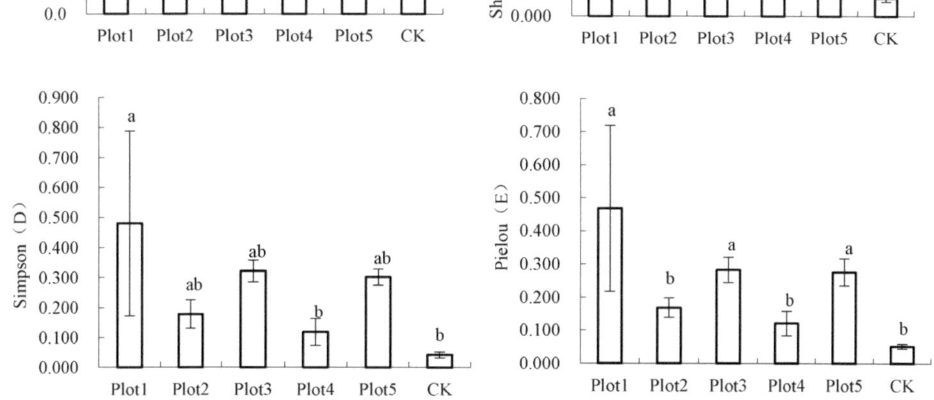

Figure 3.
Diversity of SSBs in different aged C. microphylla *plantation and control. Note: Different letters indicate statistically significant differences among plots.*

the Shannon-Wiener, Simpson, and Pielou evenness indices were also lowest in the natural grassland, (0.146 ± 0.032, 0.042 ± 0.010, 0.052 ± 0.007 respectively).

4. Discussion

One of the primary goals in restoring degraded areas is to improve species diversity and environmental stability. Consequently, much attention has been paid towards understanding the process of restoration succession and mechanisms of ecological recovery. Numerous studies have thus focused on developing effective means to rehabilitate and reconstruct ecosystems by conducting species diversity surveys in context of community recovery based on SSBs [26, 27]. In particular, the species diversity of SSB is an important component to understand restoration of degraded desert steppe environments [28]. SSB species richness and seed density are important metrics that underpin the recovery of degraded desert steppe [18]. Seed density within the topsoil (0–5 cm) was five times higher in passive restored plots than that in control treatment [29]. In this study, a total of 33 species were identified in the SSB from six plots spanning an age gradient of *C. microphylla* plantation. The increases in Gramineae grass abundances suggests improvement of grass quality that coincides with an increased species diversity in the SSB and consequently, a potentially positive effect on the recovery of ground vegetation due to the planting and maturation of *C. microphylla* plantation. Among these, annual and biennial herb species comprised 20 of the species and accounted for 58.8% of the total richness. This may be due to annual and biennial plants primarily propagated by seeds with a short life cycle, contrary to perennial plants with a longer life cycle [30]. Perennial plants commonly use asexual reproduction during drought conditions, and although they form seeds, there are far fewer than during sexual reproduction. Consequently, perennial herbs will generally occupy a smaller proportion of the SSB compared to annual and biennial herbaceous plants. A total of 35 species were identified in the aboveground vegetation, whereas 18 species were identified in the SSB and in the aboveground vegetation that included *Euphorbia humifusa*, *Setaria viridis*, *Tribulus terrestris*, *Sophora alopecuroides*, *Cynanchum komarouii*, *Corispermum mongolicum*, *Gueldenstaedtia stenophylla*, *Lespedeza potaninii*, *Euphorbia esula*, *Agriophyllum squarrosum*, *Stipa glareosa*, var. *graminifolia*, *Cuscuta chinensis*, *Bassia dasyphylla*, *Eragrostis poaeoides*, *Polygala tenuifolia*, *Artemisia scoparia*and *Salsola beticolor*. Thus, these results indicate that *C. microphylla* plantation had a significantly positive effect on the enrichment of seeds in the SSBs of the steppe. A total of 15 species only occurred in the SSB, while 17 species only occurred in the above ground vegetation, and 53% of the species were mutual to the ground vegetation and SSB. These results indicated that species composition differed between the ground vegetation and SSB communities in the Yanchi County.

Planted *C. microphylla* was correlated to a significant enrichment of seeds in the SSB. For example, the average density of seeds was 11,248.75 seeds·m^{-2} in the *C. microphylla* plantation, while the seed bank density was 661.25·m^{-2} in the semi-fixed sand dunes, representing a 17-fold decrease compared to the *C. microphylla* plantation soils. *Artemisia scoparia* seeds accounted for 82.27% - 93.81% of all seed density among plots in the different aged *C. microphylla* plantation SSB. Abundant rain lead to high levels of *Artemisia scoparia* germination in the early ground vegetation community, while late rainfall lead to the production of large amounts of seeds. These results further indicated that the SSB composition and density were related to the temporal and quantitative distribution of rainfall. The change in water spatial gradients was one of the primary factors that affected the composition, quantity, and spatial distribution of SSB, the correlation between seed banks and surface vegetation, and the seed yield of plant [31].

Seeds were significantly enriched in the 0–2 cm depth interval of the SSB of *C. microphylla* plantation, which is consistent with other studies [27, 32]. In contrast, there was a greater density of seeds below 2 cm in the semi-fixed sand dune SSB that could be due to increased porosity in the surface of the semi-fixed sand dune relative to the *C. microphylla* plantation, thereby allowing deeper movement of seeds into soils [33]. The number of seeds in the 0–2 cm SSB first increased, and then decreased with forest age, wherein seed densities followed the pattern: 25 years >16 years >7 years >37 years. This may be associated with the initial growth of *C. microphylla* plantation gradually flourishing that would help lower surface wind speed and promote the accumulation of soil nutrients, in which increased the richness of ground plant species and the seed amount in the SBB. After the growth of *C. microphylla* plantation for 20 years, the large absorption and consumption of soil water and nutrients caused the decline of the communities of other above-ground species, leading to the decrease of the number of seeds in the SBB. At present, there are few researches in this topic and the coupling analysis among SBB, aboveground vegetation and soil physical and chemical components should be strengthened in the future.

Many studies focused on analyzing SSB species diversity in order to explain the community succession and recovery mechanism, in which could be useful to guide the rehabilitation [34, 35]. Understanding the characteristics of species diversity in SSBs was an important pathway to restoration of the degraded desert steppe [36]. This study suggested that SSB diversity and evenness changed with age in *C. microphylla* plantation, which was inconsistent with the results of Bao [37]. The species density and diversity of SSB in typical steppe in Inner Mongolia of China were conducted under four season nomadic grazing (FSNG), two season rotation grazing (TSRG), settlement grazing (SG). The results of similarity between SSB and vegetation revealed that similarity coefficients under FSNG, TSRG,,FSRG and SG were 0.323, 0.351, 0.511 and 0.500 and that similarity coefficient between SSB and vegetation was higher when vegetation was more degraded [37].

The persistent seed bank only comprised three species among the plots analyzed here, indicating a lower species diversity in the persistent seed bank of desert steppe ecosystems. However, SSB species diversity was lower in the natural steppe ecosystem than in each plots of the *C. microphylla* plantation, which indicated a positive effect on SSB diversity. In order to explain the responses of SSB and vegetation to the increasing intensity of human disturbance in a semi-arid region of Northern China, four land use types were selected (native grassland, abandoned artificial grassland, artificial grassland and farmland). The results showed that native grassland had a significantly higher soil seed density and species richness than the other land use types. Moreover, the common species both in the SSB and vegetation between the native grassland and other land use types gradually decreased as disturbance intensity increased [6]. *Caragana korshinskii* of different ages was planted in degraded steppe of Mu Us Sandy Land in Ningxia. The results showed that planting *C. korshinskii* forest had a positive effect on vegetation restoration of degraded grassland. The diversity of plant community increased with the increase of the age of *C. korshinskii* [38]. Our results were similar to those of the two previous studies.

5. Conclusions

Our findings suggested that (1) the species composition of the SSB and the aboveground vegetation communities differed among *C. microphylla* plantation with different ages in Yanchi County. A total of 33 species were observed in the SSB and 35 species in the aboveground vegetation communities, moreover, 18 mutual species. Among the plots, 15 species only appeared in SSB, while 17 only appeared in the aboveground vegetation communities. (2) The *C. microphylla* plantation had a significantly positive effect on the enrichment of seeds in the SSBs. The average seed density in the SSB of *C. microphylla* plantation was 17 times than that in the semi-fixed sand dunes The SSB were primarily concentrated in the 0–2 cm soil depth interval, which accounted for 80% of the seeds in the 0–10 cm range. (3) Different aged *C. microphylla* plantation exhibited different SSB enrichments, i.e., the highest number was in 25 years plantation, then the 7 years plantation, thirdly the 16 years plantation, finally the 37 years plantation. (4) Overall, *C. microphylla* plantation resulted in positive ecological effects on the SSB species diversity in degraded desert steppe.

Acknowledgements

This work was supported by the First Class Discipline Construction Project in College of Ningxia Hui Autonomous Region (Ecology) [grant number, NXYLXK2017B06], and the National Natural Science Fund (31540007).

Author details

Guoqi Li[1]*, Shujun Li[1], Wenshan Shao[2], Yanyun Chen[1] and Wang Yafang[1]

1 Breeding Base for State Key Laboratory of Land Degradation and Ecological Restoration in Northwest China, Yinchuan, Ningxia, China

2 Forestry Technology Extension Center of Liangzhou District of Wuwei City in Gansu Province, Gansu, China

*Address all correspondence to: guoqilee@163.com

References

[1] Roberts, H.A. Seed banks in soils. Advances in Applied Biology, 1981, 6, 1-55.

[2] Alvarez-Buylla, E.R.; Martínez-Ramos, M. Seed bank versus seed rain in the regeneration of a tropical pioneer tree. 1990, Oecologia 84, 3, 314-325.

[3] Coffin, D.P.; Lauenroth, W.K. Spatial and temporal variation in the seed bank of a semiarid grassland. American Journal of Botany, 1989,76, 53-58.

[4] Johnson, M.S.; Bradshaw, A.D. Ecological principles for the restoration of disturbed and degraded land. Applied biology, 1979, 4: 141-200.

[5] Jalili, A.; Hamzeh'ee, B.; Asri, Y.; Shirvany, A.; Yazdani, S.; Khoshnevis, M.; Zarrinkamar, F.; Ghahramani, M.; Safavi, R.; Shaw, S.; Hodgson, J.; Thompson, K.M.; Pakparvar, M. Soil seed banks in the Arasbaran protected area of Iran and their significance for conservation management. Boilogical conservation, 2003, 109, 425-431.

[6] Li, C.; Xiao, B.; Wang, Q.H.; Zheng, R.L.; Wu, J.Y. Responses of soil seed bank and vegetation to the increasing intensity of human disturbance in a semi-arid region of Northern China. Sustainability, 2017, 9, 1837-1850.

[7] Defalco, L.A., Esque, T.C., Kane, J. M., Nicklas, M.B. Seed banks in a degraded desert shrubland: Influence of soil surface condition and harvester ant activity on seed abundance [J]. Journal of Arid Environments, 2009, 73(10): 885-893.

[8] Kildisheva, O.A., Erickson, T.E., Kramer, A.T., Zeldin, J., Merritt, D.J. Optimizing physiological dormancy break of understudied cold desert perennials to improve seed-based restoration. Journal of Arid Environment s, 2019, 170 :104001

[9] Yan, Q.L., Liu, Z.M., Zhu, J.J, Luo, Y. M., Wang, H.M., Jiang, D.M. Structure, Pattern and Mechanisms of Formation of Seed Banks in Sand Dune Systems in Northeastern Inner Mongolia, China. Plant and Soil volume 277, pages 175–184 (2005)

[10] Gonzalez, S.L., Ghermandi, L. Overgrazing causes a reduction in the vegetation cover and seed bank of Patagonian grasslands. Plant Soil, 2021. https://doi.org/10.1007/s11104-021-04931-y

[11] Zhao, Y., Li, M., Deng, J.Y., Wang, B.T. Afforestation affects soil seed banks by altering soil properties and understory plants on the eastern Loess Plateau, China. Ecological Indicators, 2021, 126: 107670

[12] deMedeiros-Sarmento, P.S., Ferreira, L.V., Gastauer, M. Natural regeneration triggers compositional and functional shifts in soil seed banks. The Total Environment, 2021, 753: 141934

[13] Ma, J. B., Zhang, X.M., Yan, H.L., Yang, X. L., Wang, W. H. Evaluation of Soil Seed Banks in the Taklimakan Desert. Arid Zone Research, 2010, 26 (1):71-77

[14] Shen, Y.X.; Zhao, C.Y. Soil seed bank research in China: present status, progress and challenges. Chinese Journal of Applied Ecology, 2009, 20, 467-473.

[15] Shao, W.S.; Li, G.Q.; Chen, K.Y.; Zhao, P.P. Comparative studies on soil enzymic activities of four common plant communities in desert steppe. Acta Botanica Boreali-Occidentalia Sinica, 2016, 36, 579-587.

[16] Xu, D.M.; Zhao, L.L.; Xie, Y.Z. Dynamics of plant community during sandy desertification for grassland in Yanchi County. Acta Bot. Boreal.-Occident. Sin, 2011, 31, 2084-2089.

[17] Li, G.Q.; Shao, W.S.; Zhao, P.P.; Jin, C.Q. Effects of enclosure on the soil seed bank of two plant communities on the desert steppe. Acta Prataculturae Sinica, 2018, 27, 52-61.

[18] Liu, K.; Wang, L.; Song, N.P.; Yang, X.Y.; Chen, L. Soil infiltration characteristics of Caragana korshinkii plantation with different ages in south edge of Mu Us Sandy Land. Journal of Arid Land Resources and Environment, 2013, 27, 89-94.

[19] Holmes, P.M. Depth distribution and composition of seed-banks in alien-invaded and uninvaded fynbos vegetation. Austral Ecology, 2002, 27,110-120.

[20] Yang, Y.F.; Zhu, L.; Zhang, H.Y. Analysis on the flux of soil seed bank and the seedling mortality in the communities of two species of Suaeda in the Songnen Plain China. Acta Ecoligica Sinica, 1995, 15, 66-71.

[21] Lin, J.A. A summarize of plant seed. Haerbin: Northeast forestry University Press, 1993, 211-221.

[22] Ter Heerdt, G.N.; Verweij, G.L.; Barkker, R.M.; Bakker, J.P. An improved method for seed bank analysis: seedling emergence after removing the soil by sieving. Functional Ecology 1996, 10, 144-151.

[23] Liu, R.X.; Zhan, J.; Shi, Z.H.; Chen, L. Q. Soil seed bank and its correlations with aboveground vegetation and environmental factors in water level fluctuating zone of Danjiangkou Reservoir, Central China. Chinese Journal of Applied Ecology, 2013, 24 , 801-808.

[24] Shen, Z.J.; Ou, Z.L; Tian, S.N.; Tang, W. Characters of soil seed bank in copper tailings and its adjacent habit. Acta Ecoligica Sinica, 2013, 33, 2121-2130.

[25] Zhai, F.Q.; Xu, N.; Mo, X.Q.; Li, H. Y. Characteristics of soil seed bank and soil physical and chemical properties of hydro-fluctuation belt in Jiyun Canal. Research of Environmental Sciences, 2013, 26, 97-102.

[26] Shen, Y.X; Liu, W.Y; Zhang, Y.D. The effect of rehabilitation on vegetation species composition and soil seed bank at a degraded dry valley in Dongchuan, Yunnan Province. Acta Ecoligica Sinica, 2003, 23,1454-1460.

[27] Zeng, Y.J.; Wang, Y.R.; Nan, Z.B.; Wei, D.; Chen, S.K.; Li, B.E. Soil seed bank of different grassland types of Alashan arid desert region, Inner Mongolia. Chinese Journal of Applied Ecology, 2003, 14, 1457-1463.

[28] Zhao, L.Y.; Li, Z.H.; Zhao, J.H.; Zhao, H.L.; Zhao, X.Y. Comparison on the difference in soil seed bank between grazed and enclosed grasslands in Horqin sandy land. Journal of Plant Ecology, 2006, 30, 617-623.

[29] Miao, R.H.; Song, Y.H.; Sun, Z.L.; Guo, M.X.; Zhou, Z.X.; Liu, Y.Z. Soil seed bank and plant community development in passive restoration of degraded sandy grasslands. Sustainability, 2016, 8, 581-589

[30] Hou, Z.Y; Xie, Y.H.; Yu, X.Y.; Li, F. Characteristics of soil seed bank in different water level areas after returning farmland into lake in Qingshanyuan of Dongting Lake. Chinese Journal of Applied Ecology, 2009, 20, 1323-1328.

[31] Zhu, D.G.; Ni, H.F.; Liu, F.; Wang, J.F; Cui, F.X. Quantitation characteristics of soil seed bank of deyeuxia angutifolia under different water gradient in Sanjiang plain. Heilongjiang Science, 2013, 4, 18-21.

[32] Liu, H.; Jiang, Q.; Wang, Z.J.; Pan, Z.B. A study on soil seed bank in desert steppe of different enclosing years in Ningxia. Research of Soil and Water Conservation, 2011, 18, 96-103.

[33] Dou, H.X.; Wang, M.J. The temporal and spatial dynamics of soil seed bank under different types of sand dunes in Hunshandake sandy land. Pratacultural Science, 2008, 25,116-118.

[34] Li, X.H.; Han, S.J.; Zong, W.J. Characteristics of soil seed banks of sand dune succession process in Horqin sandy land. Journal of Beijing Forestry University, 2007, 29, 66-69.

[35] Zhang, Z.T.; Song, X.Z.; Xiao, W.F. Characteristics of soil seed banks in logging gaps of forests at different succession stages in Changbai mountains. Chinese Journal of Applied Ecology, 2009, 20, 1293-1298.

[36] Tong, C.; Feng, X.; Zhang, Y.M. Soil seed banks in different grazing exclusion restoring succession stages Xilingole degraded stepps. Acta Ecoligica Sinica, 2008, 28, 1991-2002.

[37] Bao, X.X.; Yi, J.I.; Liu, S.R.; Ji, M.S.; Ji, G.J.D.S.R. Effects of different grazing system on soil seed bank in typical steppe of Mongolian plateau. Chinese Journal of Grassland, 2010, 32, 66-72.

[38] Wang, Z.J.; Jiang, Q.; Pan, Z.B.; Chen, Y.Y.; Guo, Y.Z. Changes of productivity and species diversity in resume succession on the degraded steppe of Maowusu sands in Ningxia. Pratacultural Science, 2005, 22, 5-8.

Chapter 8

Jojoba - The Gold of Desert

Raman Bala

Abstract

Jojoba [*Simmondsia chinensis* (Link) Schneider] is evergreen, perennial and drought resistant shrub belongs to the family of *Simmondsiaceae*. It is a multipurpose oil seed crop mainly grown in desert regions of world. This plant has unique oil among plant kingdom which is chemically a liquid-wax. The liquid-wax is made up of an ester of long chain fatty acids and alcohols. The liquid-wax is unique in nature because have no traces of glycerine and easily modified via hydrolysis, hydrogenation, halogenation, sulfurization, phosphosulfurization and ozonization techniques. The main uses of liquid-wax in various industries like cosmetics, pharmaceuticals, petrochemicals and lubricants. It is a potential seed oil crop for desert region so it is well known as the gold of desert. The main purpose of this chapter is to review the complete information about this plant so that it can produce and utilized maximally. Moreover, the review focuses on biology, biogeography, physico-chemical properties of jojoba oil and propagation techniques of the plant of desert regions.

Keywords: Desertification, jojoba, liquid-wax, gold of desert, propagation, unique

1. Introduction

Desertification is recognized as a major threat to global biodiversity and a burning issue to environment today. The desert land starts to degrade its chemical, physical and biological productivity. The lands get affected with a combination of various factors like deforestation, overgrazing, unsustainable agricultural practices, over exploitation of soil, poverty and climate change. Deforestation is one of them root cause of desertification [1].

India has developed Biodiversity Action Plans to counter its causes and their effect, management and conservation of endangered flora and fauna. Various environmental organizations focus on restoration of such land to control the desertification mainly in dry land areas. The NGOs focus primarily on educating the local peoples about the danger of deforestation and employ them to grow the valuable economically important plants which are easily grows in desert land.

Jojoba has very promising scope for cultivation in desert ecosystem even in the relatively hot weather. Its nature withstand in warm weather in summer and cold weather in winter at low fertility of soil. The plant need has less amount of water compared to other economic crop like olive plant with great ability to withstand in high salt in soil [2]. The commercial plantation of jojoba is encouraging mainly in Thar Desert in India. It is also known as gold of desert because it has valuable industrial application and easily grows on desert land [3].

Jojoba plantations played a vital role in combating edge effects of desertification and improve soil quality and restores land quality [4]. It is a potential oil-seed crop and helps in soil conservation, landscaping, open natural range lands and national parks. Seed hulls can be used as a mulch or protective ground cover to reduce evaporation, erosion and weed growth and can be used as fertilizers and soil replenisher for the soil which is low in organic matter. Jojoba is a valuable seed oil crop useful in preventing wind erosion in desert regions. It is already being used as an ornamental shrub throughout the southern United States [5].

Jojoba [*Simmondsia chinensis* (Link) Schneider] is a potential seed oil plant, belonging the family *Simmondsiaceae* that have played role in combating edge effects of desertification in the Thar Desert in India [6]. There are huge tracts of arid and semi arid lands in India where it has the potential to grow as an important commercial crop. Jojoba oil is unique in nature among plant kingdom. The jojoba oil has gain a lot of importance in various industries.

In 1995, an association of Rajasthan Jojoba Plantation and Research Project (AJORP) was set up by Government of India with 100% funding from the Department of Land Resource, under Ministry of Rural Development. Also, Centre for Jatropha Promotion and Biodiesel (CJP) is a global jojoba world which focuses on commercialization of entire jojoba in arid and semi-arid regions of India.

2. Biology

The other names of jojoba are hohoba, coffee berry, goat nut, deer nut, pig nut, coffee bush and lemon leaf. The first taxonomic status of this species appeared in 1822, when Johann Link named it as *Buxus chinensis* but later in 1844, it was changed to *Simmondsia californica* by Thomas Nuttall. In 1912, the Austrian botanist, Camillo Karl Schneider, renamed it as *S. chinensis* [7]. In 1897, a European botanist, Van Tieghem, proposed the creation of a new family *Simmondsiaceae*, with *Simmondsia* being the only genus containing a single species. According to the International Rules of Botanical Nomenclature, the specific name *chinensis* had first been given to jojoba that became *S. chinensis* (Link) Schneider [8].

The plant of jojoba is leafy, xerophytic, woody evergreen, dioecious in nature with small multi-stemmed that grows to a height of 0.5–1 m in the wild, occasionally to 6 m tall with taproots to 12 m long (**Figure 1a**, **b**). The natural life span appears to be over 100 years and may exceed 200 years [7].

Leaves are simple, opposite, pale green or yellowish green, oblong-ovate, rounded at both ends and leathery. The leaf apparently lives for two to three seasons, depending on moisture and shade conditions. They are 2.5 to 3.5 cm long, 0.5 to 1.5 cm wide and contain special tissue with a high concentration of phenol compounds [7].

Flowers are apetalous, unisexual and dioecious in nature. The flowers are pollinated by wind. The male flowers are yellow, larger, and occur in clusters with 10–12 stamens per flower while female flowers are small, usually solitary in the axils or in clusters at the nodes, pale green with 5 greenish sepals, soft and hairy (**Figure 1c**, **d**) [9].

It has also been reported that male plants have a smaller canopy than female plants. Plant has growing buds at axils position, actively grows and developing into a flower in the blooming season. It was noticed that these flowers buds developed at every second or third nodal position of the stem. Further, these nodal buds differentiated and grow into flowers [10].

The plant has a tap-root system. The root of mature plants can be 15–25 m below the soil surface, with substantial parallel laterals and secondary roots, giving an ability to draw moisture from the soil. Thus, allowing the plants to survive and grow where most of the other plants wither and die [11].

Figure 1.
Morphological characteristics of male, female, fruit and seeds S. chinensis (a) S. chinensis shrub (6 months) (b) mature plant (c) female flower (d) male flowers (e) immature seed (f) mature seeds.

Fruit is a green and smooth cylindrical capsule of about 2 cm length, somewhat resembling an acorn shape. The calyx of flowers is persisting with dehiscent nature and contains ovules (1–3) attached to the placenta of capsule. The capsule develops in three months and matures in five to six months. Finally, capsule splits and seeds drop outs through three sutures [8].

Seeds are peanut or acorn shaped, with a small pointed apex, and flattened base (**Figure 1e**, **f**). The color of seed is dark brown to black with smooth texture and 1–2.5 cm in long. The weight of one hundred seed can vary from 40 to 80 gm, occasionally much higher, but it is more constant for seeds from wild plants in a specific location [7].

In desert regions, plants used for revegetation as it can survive as harsh arid environmental conditions as in Rajasthan, India. The jojoba plant has potential as a soil-stabilizing nature due to its low nutrient requirements, deep root system, drought resistant, longevity and low susceptibility to fire. Jojoba is a potentially valuable industrial plant species for the control of desertification [6]. As properties oil of jojoba is same as oil of sperm whale. Therefore, more production of jojoba oil has been considerably reduced the prey of sperm whale which is now a endangered aquatic animal. The cultivation of one hectare jojoba produced oil is equal to oil derived from 124 sperm whales. Moreover, the intercropping of jojoba crop in desert land played a role of eco-friendly nature and combating desertification process [8].

3. Biogeography

The natural area of wild jojoba comprises approximately 161,000 km^2 between latitudes 23° and 34° North of latitude and between 109° and 118° West of longitudes. This shrub is native to Sonoran desert in Southern Arizona, Southern California and North-western Mexico. Now, it is being cultivated in Mexico, United States of America, Argentina, Australia, Brazil, Costa Rica, Egypt, Haiti, Israel, Paraguay and South Africa. It is used as an ornamental shrub throughout the Southern United States [7]. The plant survives in warm temperate desert to tropical desert forest. It tolerates full sun and temperatures ranging from 0 to 47°C. Mature shrubs tolerate temperatures as low as −10°C, but seedlings are sensitive to light frosts just below freezing. The rainfall 375 to 450 mm annually is considered optimum for its growth. Optimum growth occurs in a temperature range of 27–36°C, but plants can withstand very high temperature. A daily range of −1 to 50°C has been recorded in the Mexican desert habitat, but above 50°C is believed to suppress growth, although not lethal [11].

It can be grown in all types of soils which are well drained and have an average pH between 7 and 8.5. The plants has high growth in the desert soil with small amount of farm yard manure [12]. But the basic requirement is that soil must not be liable to water logging. Also, it can be grown on marginal and wasteland due to tolerance of extreme range of temperature from −5 to 54°C [13].

4. Physico-chemical properties of oil

Jojoba seeds contain about 50% oil content with a range 44 to 59% which is light gold colored and odorless liquid wax ester. Chemically, jojoba oil is a liquid wax not oil because it has neither a liquid fat nor triglycerides as all other vegetables oils. Actually, jojoba liquid wax is composed of long chain (C_{20}-C_{22}) of fatty acids and fatty alcohols with two unsaturated bonds which make the oil susceptible to many different types of manipulations. The composition of oil accounts for its extreme shelf-life stability and extraordinary resistance to high temperatures as compared with true vegetables oils. It is natural, pure and stable due to high resistance to oxidation and can be stored for years without being rancid [14].

The extracted oil is relatively pure, non-toxic, biodegradable and resistant to rancidity. It is unique among vegetable oils, as sperm whale oil is unique among animal oils [15]. Jojoba oil can be classified as non-toxic substance, and used safely for external applications on healthy skin and wounds. The physical properties of liquid-wax involve high viscosity, high flash and fire point, high dielectric constant, high oxidative stability and low volatility which makes it usable as lubricant in high pressure machinery and electric insulators. The viscosity index of jojoba oil is much higher than that of petroleum oil [7].

Jojoba oil has unusual stability, remarkably pure and need not be refined for use as transformer oil or as a lubricant for high speed machinery or machines operating at high temperatures. It used as an industrial oil for improver of lubrication power of oil of cars, airplanes and various other machines and also as a surfactant improver of paints, varnishes, plastics, waxes for better performance. Therefore, future need of huge quantities of jojoba oil for lubricant industry, especially for cars and aeroplanes but oil produced in present would not be sufficient to meet all needs of various industries.

The melting point of jojoba oil is approximately 10°C and the iodine value is nearly 80. The evaporating point of jojoba oil is 315°C which is highest in the vegetable oils. As is free from glyceride so there no need for its retiming and easily modified via hydrolysis, hydrogenation, halogenation, sulfurization, phosphosulfurization and ozonization techniques [8].

The jojoba oil does not contain sulfur therefore, mainly used in various industries sectors. The stability of oil is high and makes its very useful in the electronic and computer industries. It is also suitable for process of sulfurization to produce high quality of lubricant and rubber like material used for making ink and linoleum. Other used of jojoba oil for making plasticizers, candles, detergents and fire retardents [16].

Jojoba has potential applications in cosmetics, pharmaceutical industries, plastic industries, leather industries and bio-fuel industry. The liquid wax and its derivatives have potential in a wide range of applications in cosmetics such as lipsticks, face creams, skin fresheners, winter care lotions, soaps, shampoos and moisturizers [17].

Liquid-wax is generally used in folk remedies for renal colic, sunburn, hair loss, headache and sore throat. The oil prevents rashes and wrinkles formation on the skin if used as a night cream. It is best for hair therapy as it keeps hairs smooth and silky with a shine and also prevents hair fall-out [18].

It has wound healing properties, anti-inflammatory activity, anti-microbial activity insecticidal, antifeedant and antifungal activities. It can be classified as non-toxic substance, and used safely for external applications on healthy skin and wounds burnt skin and to remove stretch marks [19]. Also, the wax resembles human sebum and can help dry and oily skin. It penetrates into skin rapidly without leaving any traces of oil within only a few seconds as it is a straight unsaturated ester chain of fatty alcohols and fatty acids. It is used for massage because of soothing and relaxing nature and gives relief in pain of body [20].

It has no cholesterol or triglycerides and therefore can be used as low calorie edible oil [20]. Indigenous Americans and Indians used jojoba seeds and oil for cooking, hair care and for medicinal treatments such as poison ivy, sores, wounds, colds, cancer and kidney malfunction [21]. Jojoba oil, known as the sperm whale of the desert is almost identical to the whale oil, commonly known as spermaceti [22]. Jojoba oil was heralded as being able to fill this gap because of its similar chemical properties.

Jojoba also had environmental appeal as it is produced from a relatively common plant resource rather than an endangered animal species [23]. Jojoba oil is a suitable coating for some medicinal preparations, stabilizer of penicillin products and inhibitor to growth of tubercle bacilli and anti-foaming agents. It serves as cutting and grinding oil or additive to other lubricants and is suitable as transformer oil [24].

Jojoba oil is edible and act as low calorie additive for vegetable oil. It contains a chemical compound simmondsin, which depress appetite. It does not rancid and may is suitable for vegetable oil [25]. The jojoba meal extract has high 30–35% protein content and major proteins are albumins (79%) and globulins (21%) [26]. It can also be used as a component of food additive, medical food or functional food. It has been found to be useful as supplement of diet for control of body weight in humans [27].

Jojoba meal has a potential use in animal feed but needs detoxification. It has a potential use as a fertilizer because of its high nitrogen content and can be used as soil amendment to enrich soil which is low in organic matter of desert land [28].

5. Agronomy

The review of this content has led to greater understanding of the classic farming requirements for jojoba. Jojoba has the distinction of being the commercial crop successfully in desert of Rajasthan, India. To meet the demand of future its needs to be cultivated in huge amounts which are easy in desert land as well as wasteland reclamation. Soil texture is an important parameter for best growth in sandy and decomposed rocky soil. Even if the fertility of soil is marginal, jojoba is still able to produces well without use of fertilizers.

The irrigation system well established for jojoba plantation whether propagation by planting seedlings or stem cuttings. The drip watering system at root level to avoid evaporation of water so that plants use water maximally for better growth. Under ideal conditions of soil, water and sun, the tap root will grow an inch a day, within two years of planting. The root system should reach the level of aquifer thus enabling sufficient growth for seed production without supplemental watering.

Jojoba is a cross pollinated plant and crop having tendency to produce seedlings of varying size and yield. Therefore, this heterozygous progeny has raised doubt about economic feasibility of jojoba crop at commercial scale. The need of jojoba growers for the successful commercialization depends upon selection of high yields germplasm and their further multiplication [29].

Propagation of jojoba crop is through direct seeding, seedlings, rooted cutting and plantlets produced from tissue culture techniques. Commercial crop rose through direct seeds and seedlings are slow growing and requires three to four years to reach the flowering stage. These methods have limitation of male biased population which affects yields of crop [29].

When commercial plantation rose through direct seeding, about more than 50% seedlings are males. But the requirement of male population is only 10% seedlings. The plant sex can be recognized only when plants start bearing flowers after 2–3 years of planting. Therefore, a major disadvantage of plantation raised through direct seeding method. To maximize the production, plantation raised through rooted cuttings from sex-specific plants (male or female) which are known to best quality of seeds with high level of oil content. For commercial plantation with 90% female plants, leaving the 10% male plants to produces adequate pollen for all the female flowers. However, only the female plants gives the valuable seeds and requires more population in a commercial field [30].

Centre for Jatropha Promotion and Biodiesel (CJP) is a Global Jojoba World which focuses on commercialization of entire jojoba in arid and semi-arid regions of India. In Bikaner district Rajasthan of India, the commercial plantation raised through the rooted cutting of high yielding germplasm. The advantages of male and female cuttings used in a commercial field due to high uniformity of germplasm with high yield. The best planting time is spring season for well establishment of crop. Generally, jojoba crop produced an economically useful yield after four or fifth years of plantings [7].

Naturally, seed yield of jojoba range from a few seeds to as much as five kilogram seeds per plant. The seed yield varying from plant to plant and year to year production for a particular field. Average yield of seeds in a commercial plantation using selected higher yielding germplasm is capable of producing one to eleven tons per hectare after seventh to fifteenth years of planting. The irrigation system in Thar Desert of Rajasthan through the drip irrigation system supply control by pressure regulators. Irrigation system must be in initial planting time mainly in spring and early winter seasons for maximum utilization of water for growth [31].

The commercial plantations of jojoba raised successfully in Bikaner district of Rajasthan, India, shown in **Figure 2**. Naturally, flowering season of crops occurs in February to March and field image taken by author in 2019, Bikaner, Rajasthan, India (**Figure 3**). The collections of seeds manually and processed by labourers in farmhouse as shown in **Figure 4**. The jojoba meal or cake used as fertilizers in a field, image taken by author as in **Figure 5**.

Figure 2.
Well establishment of jojoba plantation in Bikaner, Rajasthan, India.

Figure 3.
Flowerings of female jojoba plant in a field of Bikaner, Rajasthan, India.

Figure 4.
Collection and processing of jojoba seeds in Bikaner, Rajasthan, India.

Figure 5.
Jojoba meal after oil removing process in farmhouse of Bikaner, Rajasthan, India.

The commercial plantation of jojoba manages by CJP through advanced crop practices adjustable to local soil quality, weather conditions, pest and disease control during frosting time. Therefore, the CJP developed a improved technology to ensures high yielding germplasm, crop management, irrigation system, fertilizers requirement, harvesting techniques, post-harvesting techniques and further grading-up process [31].

For the promotion of cultivation of jojoba as a commercial crop mainly on arid zones, the CJP organize the technical training programme on entire jojoba cultivation. The training programmes includes the selection of land sites according to soil quality and weather conditions, seed bed preparations, seeding methods, germplasm selection, planting methods, irrigation process, manure and fertilizers addition, plantation male and female ratio obtaining high quality fruits, intercropping, weeds control, diseases and their control, predators and their control, harvesting techniques, drying and storage precautions [31].

The key factors governing profitability for large scale production of jojoba that the crop is grown under such conditions which are conducive to high commercial yield. Adverse rainfall and temperature conditions can limit the commercial production of the crop. Similarly, frost is a limiting factor to reduce the quality of production. Therefore, it is evident that jojoba plantations may survive in some marginal waste land. In India, the CJP supervised the Jojoba Oil Farms to focus on the organic farming of jojoba from crop management to production of end product. Research on high yielding variety, best plantation methods with enhanced techniques has going on global scale. However, the CJP keep watching on research towards jojoba breeding techniques and collecting the useful information related to jojoba cultivation methods. The enhanced variety of jojoba with regards of seed yield and oil yield with enhanced jojoba cultivation techniques may ensures early flowering and fruiting into profitable production of oil crop.

The success of commercial cultivation of jojoba depends upon high value of market which resides on improved yielding germplasm of both male and female plants. Agriculture industry hesitant to invest in a new technology to ensures the steady and continued supply of that resource. Jojoba oil as an alternative oil in various industries and as a biofuel has been demonstrated. The jojoba crop has been proposed as gold of desert as it is highly profitable economical new oil seed crop being grown in India, Egypt, Australia and America. In India the development of jojoba cultivation including crop management, drip irrigation system and cost of production of one kilogram seeds is 0.5 $. This low cost of production in the country as compared to other oil seed crop gives it a competitive edge [31].

According to agronomist, the plantation of jojoba requires less water as compare to other oil seed crops in desert regions. This crop does not require much attention as free from cattle attacks. The economic production is good as gives high income to the farmers as compares to other traditional crop. Also, the CJP can help to farmers to identifying the needs, opportunities and solution of local, regional and national markets values. For best farming system, the CJP directs to creating best business strategy for effective cultivation of jojoba.

6. Conclusion

Jojoba is a potential seed oil crop of industrial importance mainly grows in the Thar Desert of Rajasthan, India. The arid and semi-arid land of India are all set to usher in revolution in Indian agriculture with planned large scale cultivation of the exotic jojoba plant. Jojoba in the country is now being grown on area of about thousand hectares and the area is increasing every year. About 90% of the jojoba cultivation is in Rajasthan in all over India. Jojoba bush is a valuable industrial plant for sustainable development of arid and semi-arid areas with rehabilitation of marginal land. This plant can survive in different types of soil and resist to salinity and drought conditions. Therefore, this crop has a lot of economic importance with potential to restore the land's flora and fauna, improves soil quality and prevents the process of desertification.

Author details

Raman Bala
Department of Environmental Science, Maharshi Dayanand University,
Rohtak, India

*Address all correspondence to: nandwalraman72@gmail.com

IntechOpen

References

[1] Kar, A. (2014). Desertification paradigms and measurement in Indian context. Environment, People and Development: Experiences from Desert Ecosystem (Ed.) -M.K. Gaur and P.C. Moharana. New India Publishing Agency, New Delhi, pp. 17-55.

[2] Abobatta, W.F. (2019). Overview of *Simmondsia chinensis* (Jojoba shrubs) cultivation and propagation methods. Agriculture Research and Technology, 19(2), 556089.

[3] Bala, R., Laura, J.S. & Beniwal, V.S. (2019). Direct adventitious shoot bud regeneration and *ex vitro* rooting in leaf explants of female *Simmondsia chinensis* (Link) Schneider- an important industrial shrub. Asian Journal of Microbiology Biotechnology and Environmental Science, 21, 448-454.

[4] Pasternak, D. & Schlissel, A. (2012). Combating desertification with plants. Springer science and Business Media, pp. 38.

[5] Osman, H.E. & Abohassan, A.A. (2013). Introducing Jojoba in the Arabian Desert:1. agronomic performance of nine jojoba clones selected in Makkah Area in Northern and Western Saudi Arabia. International Journal of Theoretical Applied Science, 5, 37-46.

[6] Bala, R., Laura, J.S. & Beniwal, V.S. (2020a). *In vitro* propagation of female *Simmondsia chinensis* (Link) Schneider–a Thar Desert shrub. Annals of Biology, 36, 51-55.

[7] Orwa, C., Mutua, A., Kindt, R., Jamnadass, R. & Anthony, S. (2009). Agroforestree Database: a tree reference and selection guide version 4.0 (http://www.worldagroforestry.org/sites/treedbs/treedatabases.asp)

[8] Thomson, P.H. (1982). Jojoba Handbook (3rd Edition). Bonsall Publications, 4339 Holly Lane Bonsall, California 92003, USA.

[9] Gentry, H.S. (1958). The natural history of Jojoba (*Simmondsia chinensis*) and its cultural aspects. Economic Botany, 12, 261-295.

[10] Botti, C., Franck, N., Prat, L., Carrasco, O. & Canaves, L. (2001). Caracteristicas botanicas. In: La Jojoba (*Simmondsia chinensis* (Link) Schneider). Serie Ciencias Agronomicas No. 5 Facultad de Ciencias Agrarias. Universidad de Chile, Santiago, pp. 5.

[11] Weiss, E.A. (1983). Crambe, niger and jojoba. In: Oilseed Crops. Longman, London, UK, pp. 507-527.

[12] Bala, R. & Laura, J.S. (2015). Effect of various soil medium on seed germination of *Simmondsia chinensis* (Link) Schneider (jojoba) under *in vivo* conditions. International Journal of Pharm and Biological Science, 6(B), 77-81.

[13] Hall, N. (1996). Mallee Gross Margins, 1996-97, Agriculture Victoria, Swan Hill, Victoria.

[14] Zaher, F.A., El Kinawy, O.S. & El-Haron, D.E. (2004). Solvent extraction of jojoba oil from pre-pressed jojoba meal. Grass Aceites, 55, 129-134.

[15] Bala, R. & Laura, J.S. (2013). Micropropagation of *Simmondsia chinensis* (link) Schneider: a review on the regeneration potential of different explants. International Journal of Current Research, 5, 3610-3617.

[16] Agrawal, V., Sharma, K., Gupta, S., Kumar, R. & Prasad, M. (2007). Identification of sex in *Simmondsia chinensis* (jojoba) using RAPD markers. Plant Biotechnology Reports, 1, 207-210.

[17] Reddy, M.P. & Chikara, J. (2010). Biotechnology advances in jojoba

(*Simmondsia chinensis*). In: Ramawat, K.G. (Eds.). Desert plants: Biology and Biotechnology, Springer-Verlag, Berlin, Heidelberg, pp. 407-422.

[18] Ranzato, E., Martinotti, S. & Burlando, B. (2011). Wound healing properties of jojoba liquid wax: an *in vitro* study. Journal of Ethnopharmacology, 134, 443-449.

[19] Guirguis, O.W., Abd Elkader, M.F.H. & Nasrat, A.A. (2013). Enhancing antimicrobial activity for chitosan by adding jojoba liquid wax. Material Letters, 93, 353-355.

[20] Undersander, D.J., Oelke, E.A., Kaminski, A.R., Doll, J.D., Putnam, D.H., Combs, S.M. & Hanson, C.A. (1990). Jojoba: Alternative field crops manual, University of Wisconsin-Exension, Cooperative Extension.

[21] Agrawal, V., Prakash, S. & Gupta, S.C. (2002). Effective protocol for *in vitro* shoot production through nodal explants of *Simmondsia chinensis*. Biologia Plantarum, 45, 449-453.

[22] Sharma, S.K. & Singh, A.P. (2011). Pharmacognostical evaluation of roots of *Simmondsia chinensis* (Link) Schneider. International Journal of Pharmaceutical Science and Drug Research, 3, 323-326.

[23] Morrissey, H., Van, R., Waston, B. & Oliver, M. (1996). Jojoba: Farm Level Analysis and Market Assessment. ABARE report prepared for the Murray-Darling Basin Commissions.

[24] Yermanos, D.M. (1982). Jojoba- a potentially valuable species in the control of desertification. Proceedings of the Conference on Alternative Strategies for Desert Development and Management, 31 May-10 June 1977, United Nations Institute for Training and Research, Sacramento, California, USA. Agriculture, 2, pp. 374-381.

[25] Aftab, F., Akram, S. & Iqbal, J. (2008). Estimation of fixed oils from various explants and *in vitro* callus cultures of jojoba (*Simmondsia chinensis*). Pak Journal of Botany, 40, 1467-1471.

[26] Shrestha, M.K., Peri, I., Smirnoff, P., Birk, Y. & Golan-Goldhirsh, A. (2002). Jojoba seed meal proteins associated with proteolytic and protease inhibitory activities. Journal of Agriculture Food Chemistry, 50, 5670-5675.

[27] Teague, R.K., Tynch, S.L., Jaksch, F.L. & Maier, R.T. (2005). Compounds for altering food intake in humans. United States patent 6852342. http://patents.Justia.com/06852342.

[28] Hosseini, F.S., Hassani, H.S., Arvin, M.J., Baghizadeh, A. & Nejad, G.M. (2011). Sex determination of jojoba (*Simmondsia chinensis* cv. Arizona) by random amplified polymorphic DNA (RAPD) molecular markers. African Journal of Biotechnology, 10, 470-474.

[29] Al-Obaidi, J.R., Suliman, B.K. & Al-Ani, N.K. (2012). Molecular-based marker for sex determination of jojoba *in vivo* and *in vitro* Iraqi cultivats RAPD-PCR technique. Scientific Research and Essays, 7, 522-527.

[30] Bala, R., Laura, J.S. & Beniwal, V.S. (2020b). Large scale production of female plants of *Simmondsia chinensis* (Link) Schneider through synthetic seed technology. Plant Cell Biotechnology and Molecular Biology, 21, 47-53.

[31] Jojoba: A potential tree borne oil seed. Jatropha World-Generating green power for people and planet. https://www.jatrophabiodiesel.org/Jojoba/index.php

Section 3

Desertification Control

Chapter 9

Role of Eco-Village Initiatives in Mitigating Desertification in Semi-Arid Areas of Tanzania

Fredy S. Mswima and Abiud L. Kaswamila

Abstract

Climate change adaptation actions for mitigating desertification and improving community livelihood in developing countries have attracted numerous scholarly works. However, there have been insufficient findings on the adaptation regarding the eco-village practices in semi-arid areas in particular. This inspired a study to assess the role of eco-village practices in strengthening climate change adaptive capacity and mitigating desertification in semi-arid areas of Chololo village, Dodoma region in central Tanzania. Data were collected using mixed methods, that is, household survey (92), focus group discussions (21), key informants interviews (6), field observation and documentary review. Statistical Package for Social Sciences (SPSS) and content analysis were used in analyzing quantitative and qualitative data respectively. The study found a relatively high level of community awareness on the eco-village initiative; the initiative rehabilitated village forest reserve; improved land productivity for sorghum and pearl millet; increased number of planted trees; and strengthening communities' adaptation to climate change through improved households' nutrition, income and reduced water stress.

Keywords: eco-village, Chololo, Dodoma, Tanzania, desertification, climate change, semi-arid

1. Introduction

The growing global climate change has already had observable effects on the environment and human welfare [1–4]. The impact has been manifested through changes in global average temperature that has increased by 0.8-1°C over the past 100 years and 2016 was the hottest (0.99°C) year on record [5]. Rainfall models indicate increases of precipitation near the equators, Arctic and Antarctic. The Mediterranean and Southern Africa regions have experienced precipitation drop of about 20%. Western Australia, Chile, and Central America/Mexico are likely to become around 10% drier [6]. Globally, these climate change indicators have had implications to the increased type, frequencies and intensities of extreme weather events like floods, droughts, tropical cyclones (hurricanes and typhoons) and heavy precipitation [7]. The growing impact has widely affected agriculture, water, livestock, forestry and the general ecological systems mostly in developing countries [8].

The climate change impacts in developing countries have mostly been noted through extended periods of droughts, loss of soil fertility, shortening of the growing season negatively affecting crop yield that in turn worsen food insecurity, land degradation and desertification and subjecting many people at risk of hunger. As for the forestry sector which is resorted by more than 90% for wood fuel in Africa and Asia has recorded severe biomass deterioration through loss of biodiversity, limited forest products, ecosystem shift from forest to woodlands or woodland to grasslands and desertification [8].

The impact of climate change has compelled several interventions to redress the effects particularly in semi-arid areas of developing countries including use of Clean Development Mechanisms (CDM) and Reducing Emissions from Deforestation and forest Degradation (REDD) that aimed at commodifying forestry conservations efforts to the community owning and/or living adjacent to the managed forest(s). However, it has been reported that the initiatives inadequately embraced the socio-economic welfares of the community that were claimed to be the drivers for the deteriorating forests in the course of earning their livelihoods [9]. REDD, in particular, unsuccessfully supported alternative activities such as crop farming, domestic energy, beekeeping and brick making as means to redress shifting agriculture, exploitation of wood fuel and encroachment that were regarded as underlying drivers of deforestation and forest degradation [10].

Despite the ongoing debates on the effectiveness of CDM and REDD, the countries severely hit by the impact of climate change have embarked on the eco-village initiative as one of the renewed paradigm shifts to directly strengthen capacity of community for adaptation against negative impact of climate change. In the context of this study, eco-village means rural communities/settlements/village in a landscape devoted to increasing resilience and capacity to adapt to climate change through an integrated and multidisciplinary range of Climate Change Adaptation (CCA) activities in an indefinite future.

The eco-village initiative is an integral part for enabling community to adjust to the climate change focusing on consolidating people's welfare sectors, namely crop farming, livestock, forestry and ecological systems [11, 12]. The objective of this is study is to assess the impact of the eco-village initiatives and/or practices in mitigating climate change and desertification in semi-arid areas of central Tanzania. Specifically, the chapter intends to (i) t**o identify eco**-village initiatives and/or practices for Adaptation to Climate Change and mitigation of desertification and (ii) to assess the performance of eco-village initiatives and/or practices in smallholder farmer's yields before and after the eco-village interventions.

2. Methodology

2.1 The study area

The study was carried in Chololo village (**Figure 1**) located in Dodoma City one of the semi-arid areas in Tanzania. Dodoma is located at Coordinates 6°57′ and 3°82' South of the Equator and between longitudes 36°26′ and 35°26′ East of Greenwich in the center of Tanzania. Chololo village has six sub-administrative villages, namely Kawawa, Lusinde, Jamhuri, Muungano, Siasa and Kizota. According to the United Republic of Tanzania 2012 census, Chololo village had 1111 households [13]. The site was previously characterized with vulnerable and deprived farming families dominated by the Gogo ethnic group, predominantly engaged with rain-fed agriculture and subsistence farming. Semi-arid areas is one of the agro-ecological zones of Tanzania distinguished by the low to medium undulating plains (200–1,500 m asl)

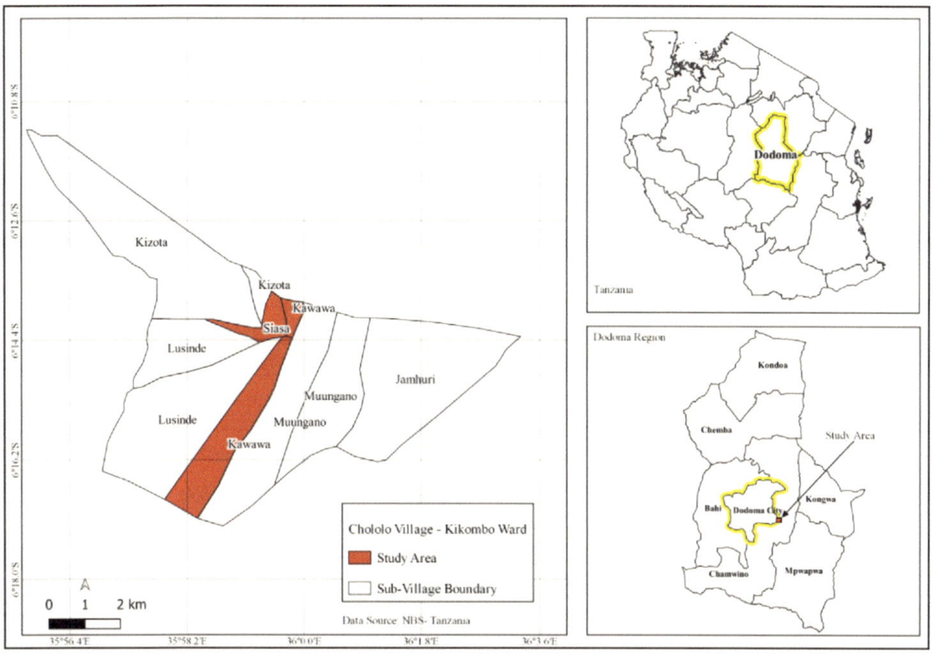

Figure 1.
Map of the study area.

with rocky hills and low scarps in central and south-eastern areas, with soils of variable fertility, localized salinity and hard-pan problems, and unreliable unimodal rainfall (500–800 mm per year) [14].

2.2 Data collection methods and analysis

Mixed research methods – household survey, key informant interviews, Focus group discussion, physical visits and documentary review were used in the study. Household Survey involved collection of both qualitative and quantitative data using a pre-designed household questionnaire. Key informant interviews involving village officials and village subject matter specialists were also planned. FGDs involving the elderly, youths and females were also held. Other methods were physical visits so as to assess the situation on the ground in crop farming, livestock, water resources, and forestry sectors and documentary reviews for the purpose of supplementing the collected field data.

The study employed both probability and non-probability sampling. Purposive sampling was used in selecting household heads, key informants, and focus group discussions participants. Purposive sampling is mostly applicable in qualitative research in which the representative population is identified and selected to provide rich information using limited resources [15]. In addition, it is useful in identifying and selecting the knowledge rich and experienced individuals regarding the phenomenon of interest [15], Simple random sampling using standard procedures was adopted in selecting two village hamlets (Kawawa and Siasa) out of four (Lusinde, Kawawa, Jamhuri and Kizota). Simple random sampling design renders equal chance of individuals in being picked up to form a sample; it is bias-free, time-efficient and cost effective; and data from this design can easily be analyzed and it is possible to generalize the results [16]. Statistical Package for Social Sciences (SPSS) and content analysis were used in analyzing quantitative and qualitative data respectively.

3. Results and discussion

3.1 Socio-economic characteristics of the communities

Socio-economic characteristics are presented in **Table 1** below. Results indicate that, on average, males were 83.7% of the respondents. This could have happened by chance. In terms of age, 54.3% were adults aged between 40 and 55 years old and 19.6% were youths aged between 24 and 39 years. Others (26.1%) were elders aged more than 55 years old. The age groups mean that the majority were economically active. In terms of education, more than 76.2% had primary education, implying that the illiteracy level was still high. Regarding household size, 51.2% had more than 4 people, 42.3% comprised of 3–4 people and 6.5% had 1–2 people. Findings further indicate that area is dominated by Gogo ethnic group and the majority (74%) practice both agriculture and livestock production. Only 26% depend solely on agriculture for a living.

3.2 Eco-village practices for adaptation to climate change and mitigating desertification

The common eco-village practices identified through thematic analysis are presented in **Figure 2** below. Of the several eco-village practices and/or initiatives use of oxen-drawn techniques, rehabilitation of village forest, improved chicken/ poultry, use of energy saving stoves and afforestation were ranked by respondents as the most common.

Socio-economic characteristic variables		%
Sex of the household head	Male	83.7
	Female	16.3
Age	24–39	19.6
	40–55	54.3
	>55	26.1
Marital status	Unmarried	5.5
	Married	82.6
	Separated	3.2
	Divorced	—
	Widow(er)	8.7
Education	Primary	76.2
	Secondary	8.7
	Informal	15.1
Household size (%)	1–2 people	6.5
	3–4 people	42.3
	>4 people	51.2
Household economic activity	Crop farming & livestock/poultry	74
	Crop farming only	26
Ethnic group	Gogo	100

Source: Field data (2017).

Table 1.
Socio-economic characteristics.

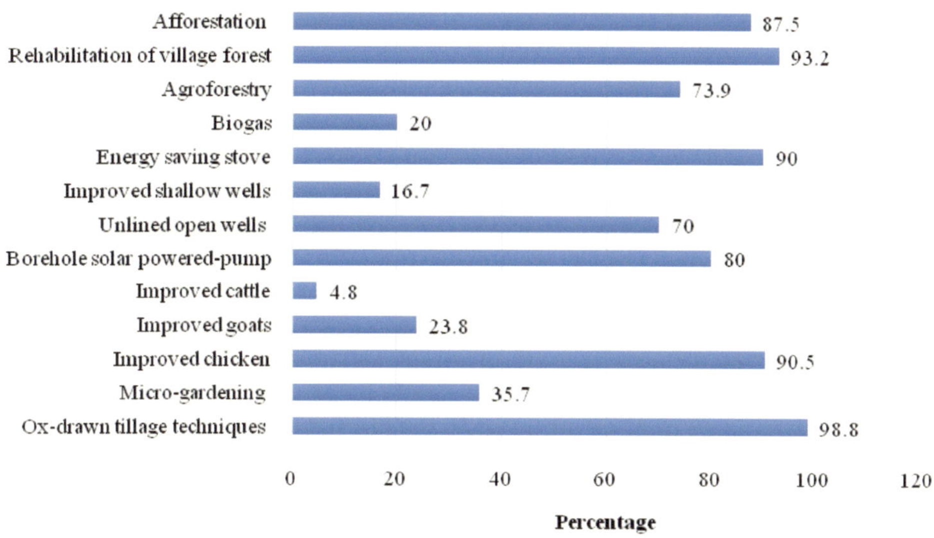

Figure 2.
Major eco-village practices in Chololo eco-village. Source: Field data (2017).

The above five initiatives can be grouped into two major categories. First, oxen-drawn techniques and improved chicken/poultry initiatives were meant for livelihood. The fact that agriculture is the lifeline of these small-holder farmers use oxen to cultivate instead of hand-hoe hoping to make a difference in terms of increased plot size and land preparation hence increased production of drought resistant crops in the area, that is, sorghum, pearl millet, sunflower and groundnuts and improving soil structure.

During interview with one eco-village official in Chololo eco-village, he argued that turning over the upper soil with oxen-plow was advantageous in bringing fresh nutrients to the surface while burying weeds and previous crop residues and/or manure and allows decay of organic matter. Similarly, community involvement in improved chicken/poultry has been used as source of income and in improving household nutrition.

Second, rehabilitation of village forest, use of energy saving stoves and tree planning are meant to reduce forest and soil degradation, protection of water sources and protection of soil cover which in turn can play a very big role in mitigating desertification. Furthermore, energy saving stoves was meant to provide domestic energy for cooking and heating and indirectly spare much of forestry resources. Scholars (e.g. [17–21] suggest that the adverse impacts of climate change, appropriate responses and adaptive capacity advocated by the eco-village initiative are relatively more site-specific.

3.2.1 Oxen-drawn tillage

We used Binary Logistic Regression Analysis to assess community's motive on oxen-drawn tillage techniques and results shows that the estimated odds ratio was 5.62 indicating that the households' cropland preparation using oxen-drawn tillage techniques was 5.62 more appropriate for healthier crop growth in response to the impact of climate change compared to those who never used the techniques. Oxen-drawn tillage were appropriate techniques in response to the growing impact of climate change manifested through unpredictable rainfall, declining soil fertility and abnormal loss of soil moisture content that lowered cropland productivity.

The appropriateness of ox-drawn tillage technique in terms of user-friendliness and fitting to local knowledge and skills for generation of high benefits is consistent with the innovation-diffusion theory [22] which emphasizes that relevance of the practice motivates the users in carrying it for realization of benefits.

Views by eco-village initiative official, FGD participants and community seed producers in the study area had the opinions that the increased land productivity was a result of abandonment of previous slash-and-burn land preparation method, which was common in the past and adoption of oxen drawn tillage practice. Slash-and-burn used to remove grasses and any remaining crop residues. Despite slash-and-burn minimal effect on soil disturbance, still the unbroken hardpans restricted drainage of water and growth of plant roots in the face of growing impact of climate change. This decreased crop productivity and increased food insecurity and resulted into the increase in the number of people at risk from hunger in the area.

Oxen-drawn tillage involves turning over the soil and loosening it (**Figure 3**) immediately after the first rain season for sowing the early maturing and drought tolerant sorghum, pearl millet, sunflower and groundnuts, normally between late December and mid-January. During the interview with the eco-village official in Chololo eco-village, he argued that turning over the upper soil was advantageous in bringing fresh nutrients to the surface while burying weeds and previous crop residues and/or manure and allowing decay of organic matter.

Compared to slash-and-burn, ox-tilled land was useful in promoting soil moisture and fertility, and in reducing soil erosion; which in turn promote crop growth in a limited rainfall caused by the impact of climate change. The use of ox-drawn tillage in semi-arid areas was found to promote soil fertility, reduced soil erosion, and increased soil capacity in retaining rainwater [23, 24]. The ox-drawn tillage technique was of paramount importance adaptation practice in place against the unpredictable rainfall, the ever-declining soil fertility, increasing soil erosion and loss of moisture content through run-off and evaporation, which are mostly accelerated by climate change.

3.2.2 Rehabilitation of village Forest

Forest rehabilitation can be defined as the process of restoring the capacity of the degraded forest land to deliver forestry products and services [25]. In the context of this study, rehabilitation refers to the re-planting of degraded village trees in

Figure 3.
Farmer in Chololo eco-village cultivating using oxen-drawn plow in December 2017. Source: Field data (2017).

the village forest reserve to re-establish the presumed forest structure, productivity and tree species diversity for future delivery of the desired products and services.

We used Binary Logistic Regression Analysis to assess community's motive on rehabilitation of the Village Forest Reserve. Findings show that the rehabilitation of village forest reserve advocated by the eco-village initiative was 42 times more compatible with culture of restoring forest services to the participants in the era of increasing climate change stress to nearby forests compared to the non-participants. This means that participants of tree planting were more likely to restore the negative effects of over-exploitation of the long-standing village forest reserve for several uses such as wood fuel for domestic uses and extraction of building materials. Drawing from the adoption perception theory, the compatibility of the practice with the individuals' previous behavior, practice and/or way of living (for this case tree planting) is one of the motives for participation in the practice [26].

The loss of such previously vegetated forests had bearing on the increased rate of soil erosion, loss of soil fertility and wind disturbances. Similarly, the interview with one of the participants in tree planting in Chololo showed that, initially, before the eco-village initiative, forests management and conservation were actively observed; but deteriorated with time to respond to the impact of climate change manifested through elevated soil erosion and poor forest related services. In addition, forest resources became inadequate to support the growing wood fuel and building/construction demands.

The interviews with key informants and all FGDs showed that, after being mobilized and sensitized on environmental education and along with initial facilitation with seedling, the villagers participated in re-planting of such trees as cutch (*Acacia polyacantha*) and neem (*Azadrachita indica*) in the degraded village forest reserve. Besides facilitation of community members and village leaders on afforestation, nursery management and tree planting in Chololo, it was observed that community were involved in planting about 3000 trees within the village forest reserve [24].

The practice of rehabilitation of village forestry is in line with the Tanzania's environmental policies that call for immediate measures in restoring the degraded forests to assume the productive state, partly promoting both environmental goods and services for the community [27]. The community's venture on rehabilitation of village forest reserve partly reflects the community's commitment in ensuring sustenance of village forestry services like building materials, windbreak, reduced erosion, scenic value in the course of adapting to climate change and mitigation against land degradation and desertification.

3.3 The performance of eco-village practices for adaptation to climate change and mitigation of desertification

3.3.1 Changes in average of small-holder farmers' yield

Paired-samples t-test was conducted to compare the average small-holder farmers' yield per hectare for four main crops - sorghum, pearl millet, groundnuts and sunflower before eco-village intervention and after intervention. Results indicates that there was a significant difference in the average small-holder farmers' yield per hectare in sorghum and pearl millet ($t = 5.361$, $p < 0.05$) and ($t = 6.656$, $p < 0.05$) respectively, compared to the situation before intervention (without using oxen). On the contrary, the change in average small-holder farmers' yield for sunflower and groundnuts ($t = 1.893$, $p > 0.05$) and ($t = 1.338$, $p > 0.05$) was statistically insignificant (**Table 2**).

An important implication of significant average increase in yield per hectare for sorghum and pearl millet is probably due to the fact that such crops are both major

Paired Samples Test											
			Paired Differences					t	df	Sig. (2-tailed)
			Mean	Std. Deviation	Std. Error Mean	95% Confidence Interval of the Difference				
						Lower	Upper			
Sorghum	Pair 1	YA- YB	.36	.44	.07	.22	.49	5.36	42	.00
Pearl millet	Pair 1	YA- YB	.42	.36	.06	.29	.55	6.66	32	.00
Sunflower	Pair 1	YA- YB	.34	.79	.18	−.04	.72	1.89	18	.07
Groundnuts	Pair 1	YA- YB	.13	.43	.09	−.07	.32	1.34	20	.19

YA, Average yield after the eco-village intervention; YB, Average farmers' yield before the eco-village intervention.
Source: Survey data (2017).

Table 2.
Changes in average yield for the major crops in Chololo.

staple food and cash crops and increased production creating surpluses motivated small-holder farmers to devote more efforts to transform knowledge and skills to increase land productivity. Increase in sorghum and pearl millet was also supported by key informants and FGD participants in the study area who argued that "Oxen-drawn tillage encouraged use of proper agronomical practices like appropriate spacing, thinning and weeding eventually, increased the number of meals taken per day from one to two". It was further emphasized that ox-drawn tillage techniques promoted soil moisture and fertility, and reduced soil erosion and compaction; thus boosting the survival rate of the drought tolerant crops, high yielding and early maturing crops like sorghum and pearl millet. Eventually, it increased the land capacity in sustaining the crop growth under poor rainfall resulting from climate change.

The studies by [21, 24], in semi-arid areas reported the growing rate of small-holder farmers' abandoning the traditionally low crop yield slash-and-burn (clearing land and sowing) to modern ox-drawn tillage (deep plowing and sowing) techniques especially after the eco-village initiative [24] also found that small-holder farmers in semi-arid areas managed to increase crop yield per cropland size cultivated by valuing and translating the acquired knowledge on ox-drawn tillage land preparation method into practice. As a result, there was improved household food security manifested in limited migration of community in search of foods, increase in number of meals to about two-three per day, and the reduction in number of months without food [21]. Drawing on the economic constraints theory, [28] argue that the decision made by the user/adopter of practice is consolidated by the experienced yields and profits.

3.3.2 The gap between the potential yield and average smallholder small-holder farmers' yield

Table 3 presents results on the average gap between potential and average small-holder farmers' yields for major crops cultivated in Chololo eco-village. The gap between potential and average small-holder farmers' yields for major crops in Chololo ranged from 51.5% for pearl millet to 55.2% for sorghum after the initiative as of before ranged from 68.5% for pearl millet to 69% for sorghum.

In filling the gap, several scholars suggest a number of measures to crop production: firstly, by timely applications of irrigation and fertilizers based on crop requirements and soil conditions [29]. Secondly, by promoting smallholder farmers' use of water-saving techniques such as mulching [30] and sprinkler irrigation

Major crop type	Performance before 2011 (at start)				Performance after (6 years)			
	FY (t/ha)	PY (t/ha)	Change (PY-FY)	% change	FY (t/ha)	PY (t/ha)	Change (PY-FY)	% change
Sorghum (*Sorghum bicolor*)	0.85	2.7	1.85	68.5	1.21	2.7	1.49	55.2
Pearl millet (*Pennisetum americanum*)	0.62	2	1.38	69	0.97	2	1.03	51.5

FY, Average small-holder farmers' yield; PY, Potential Yield.
Source: Survey data (2017).

Table 3.
Average small-holder farmers' yield in relation to potential yield.

(instead of furrow irrigation) to deal with the increasing water shortages in most of the areas. Thirdly, by improving the agro-technical service provision and government support would likely help smallholder farmers adjust their crop management and adopt more sustainable farming practices. Fourthly, by including the development of policies associated with land tenure to stimulate cooperation of individual households or allowing land transfer to merge the very small and fragmented farms (often less than one hectare) into larger land holdings. This would increase the yields in terms of efficiency use of labor and other production factors such as machinery, resulting in low labour productivity and profit [31].

The untimely availability or completely lack of industrial fertilizer and/or manure made small-holder farmers miss the correct application of fertilizer on crops to boost cropland productivity. Lack of oxen and plow, delayed some small-holder farmers in catching up with farm preparation and planting. Meanwhile, the repeated rainfall unpredictability and poor distribution in some incidences, impaired the whole crop farming planning processes resulting into massive crop failures, thus reduced crop yields.

Limited knowledge and lack of skills in crop production among small-holder farmers constrained crop production too; for instance, through inefficiencies in meeting the desired conditions of practices and failure to make appropriate choice of seeds. Similarly, previous studies point out that the shortage of farm implements; inadequate education on crop farming practices [30] and erratic rainfall on the rain-fed crop farming [31] are among the major factors that affect crop productivity in crop farming.

To promote the performance of such practices of the eco-village under crop farming, several suggestions were pointed out, namely: the government and all relevant stakeholders should participate effectively in timely provision of farm implements; they should promote knowledge and skills on crop farming through workshops, seminars, farmer field schools and exhibitions; and they should promote the user-friendly weather forecasting and prediction, and disseminate user-friendly information to the grass-roots levels involved in crop farming in rural areas.

3.3.3 The survival rate of trees in the Rehabilitated Village Forest reserve

A paired-samples t-test was conducted to compare the average tree survival after the rehabilitation of the village forest reserve advocated by the eco-village initiative and planted trees in village forest before intervention. Results showed that there was no significant difference in the average survival rate of trees in the rehabilitated village forest reserve ($t = -1.41$, $p > 0.05$) compared to the situation with no eco-village intervention (**Table 4**). The associated tree planting limitations in terms of costs, efforts, and maintenance can explain this.

		Paired Differences					t	df	Sig. (2-tailed)
		Mean	Std. Deviation	Std. Error Mean	95% Confidence Interval of the Difference				
					Lower	Upper			
Pair 1	SA-SB	−1.20	2.69	.85	−3.13	.73	−1.41	9	.19

SA, Average number of tree survival before; SB, Average number of tree survival after eco-village intervention.
Source: Survey data (2017).

Table 4.
Average number of survived trees.

Furthermore, the slow growth rates and limited direct linkage of forest resources to the urgent felt needs can have bearing on the reduced community's efforts in maintaining planted trees against harsh conditions and environment. One of the respondents and the eco-village official emphasized that unrestricted animal movements and untimely replacement of dead trees undermined the survival of planted trees in the area. On the other hand, focus group discussion for adult females had the opinion that "scarcity of water posed risk to survival of trees in the area". On the other hand, adult female focus group discussion had the opinion that "scarcity of water posed risk to the survival of the trees progress in the area" [21] observed water scarcity and limited urgent response of tree planting practices to livelihood felt needs led to low survival of the trees planted. Other causes raised by respondents were inadequate supply of quality seedlings which demoralized the community's participation in tree planting at household level and effect of termites on the transplanted seedlings during dry season [30, 31]. Reports that tree pests as well as insects and fungi, limited watering, and unavailability of quality seedling is among the factors hindering tree planting programmes at household and at the institutional levels.

To redress the challenges, the study suggests timing of planting trees during the wet season, establishment of charcoal dams for irrigating trees in areas with unreliable sources of water and selection of seedling with high survival rate under minimal watering. Meanwhile, the shortage of seedling can be addressed by promotion of seedlings at the local level that can be established and managed by individuals, groups and institutions like schools and the village for sustainable tree planting programme. In controlling termites, the study suggests the use of pesticides and/or adequate watering to manage soil wetness so as to boost the growth of transplanted seedlings at early stages.

4. Conclusions

The community in semi-arid areas of Chololo eco-village carried out practices that had a direct response to livelihood and success in addressing the impact of climate change and desertification. The practices and/or initiatives mainly, included oxen-drawn tillage and rehabilitation of village forests. These initiatives seemed to be appropriate in enhancing cropland productivity through reduced soil erosion, improved soil fertility/productivity and moisture conservation; improvement of income and food security; and mitigating land degradation and desertification. Despite the positive contribution of the crop farming practices in lessening desertification and improved livelihood, we noted that the average farmers yield were below the potential yield. The survival rate of trees with regard to forestry rehabilitation was not encouraging to attain desirable effects in mitigating desertification. It is recommended that the local communities in the study area, village institutions and other relevant stakeholders should promote the best performing eco-village practices that have been evolved through participatory approaches; and Dodoma City Council should regularly facilitate the communities through provision of knowledge and technical support to sustain the eco-village practices.

Acknowledgements

We extend our appreciation to the University of Dodoma and the European Union for funding this study.

Author details

Fredy S. Mswima and Abiud L. Kaswamila*
Department of Geography and Environmental Studies, The University of Dodoma, Dodoma, Tanzania

*Address all correspondence to: abagore.kaswamila6@gmail.com

References

[1] Breidenich, C., Magraw, D., Rowley, A., & Rubin, J. W. (1998). The Kyoto protocol to the United Nations framework convention on climate change. American Journal of International Law, *92*(2), 315-331.

[2] Solomon, S., Qin, D., Manning, M., Chen, Z., Marquis, M., Averyt, K. B., ... Miller, H. L. (2007). Contribution of working group I to the fourth assessment report of the intergovernmental panel on climate change, 2007: Cambridge University Press, Cambridge.

[3] Stern, N., Peters, S., Bakhshi, V., Bowen, A., Cameron, C., Catovsky, S., ... Edmonson, N. (2006). *Stern Review: The economics of climate change* (Vol. 30): HM treasury London.

[4] UNFCCC. (1992). United Nations framework convention on climate change: United Nations New York.

[5] Osborn, T., Barichivich, J., Harris, I., van der Schrier, G., & Jones, P. (2017). Monitoring global drought using the self-calibrating Palmer Drought Severity Index [in" State of the Climate in 2016"]. Bulletin of the American Meteorological Society, *98*(8), S32-S33.

[6] Hausfather, Z. (2018). What climate models tell us about future rainfall, Published by Carbon Brief Ltd - Company No. 07222041.

[7] Meehl, G. A., Stocker, T. F., Collins, W. D., Friedlingstein, P., Gaye, T., Gregory, J. M., ... Noda, A. (2007). Global climate projections.

[8] McCarthy, J. J., Canziani, O. F., Leary, N. A., Dokken, D. J., & White, K. S. (2001). *Climate change 2001: impacts, adaptation, and vulnerability: contribution of Working Group II to the third assessment report of the Intergovernmental Panel on Climate Change* (Vol. 2): Cambridge University Press.

[9] URT. (2011). *Tanzania Agriculture and Food Security Action Plan, 2011-12 to 2020-21*. Retrieved from Dar es salaam, Tanzania:

[10] Mswima, F. S., & Kanyama, A. (2014). The Implication of livelihood situation of local community on Implementation of REDD+ project in Kilosa District, Tanzania.

[11] Danielsen, F., Skutsch, M., Burgess, N. D., Jensen, P. M., Andrianandrasana, H., Karky, B., ... Ngaga, Y. (2011). At the heart of REDD+: a role for local people in monitoring forests? Conservation letters, *4*(2), 158-167.

[12] Kimbowa, R., Mwayafu, D., & Njaidi, R. (2011). Benefit sharing to make REDD+ work for communities and forest conservation in Tanzania: The Community Carbon Enterprise (CCE) Model. *REDD-net programme*.

[13] URT. (2012). *Monitoring and Evaluation Framework for Climate Change Adaptation in Tanzania*. Dar Es Salaam, Tanzania: Vice P

[14] URT. (2014). *State of the Environment Report 2*. Dar Es salaam, Tanzania: The Vice President's Office, Environment.

[15] Engelbrecht, M. N. (2015). *The effect of supplements containing different protein and energy sources and essential oils on the performance of pasture finished heifers*. Stellenbosch: Stellenbosch University.

[16] Garnett, J. (2016). Grassroots Environmental Adult Education: Developing Environmental Peace Infrastructure in the Nascent Democracy of Myanmar. Peace & Conflict Review, *9*(1).

[17] Mdoe, S. Y., & Kilawi, C. J. (2014). *Enhancing Climate Change Adaptation and Ecosystems of the Uluguru Mountains,*

Final Project Evaluation Report: Sokoine University of Agriculture.

[18] Myles, K., Olesen, G. B., Tamilselvam, D., & Poonia, A. (2015). *"Evidence based advocacy for low-carbon, pro-poor sustainable "Eco-Village Development" (EVD) in South Asia*. Retrieved from

[19] Njau, B. F., Farrelly, M., & Msumbuko, I. (2014). Climate Vulnerability and Capacity Analysis in Dodoma Region of Tanzania, F. Bernard Njau, Research Desk, 2014, . *3(1)*, 374-391.

[20] Gebregziabher, S., Mouazen, A. M., Van Brussel, H., Ramon, H., Nyssen, J., Verplancke, H., ... De Baerdemaeker, J. (2006). Animal drawn tillage, the Ethiopian ard plough, maresha: a review. Soil and Tillage Research, *89*(2), 129-143.

[21] ESRF. (2014). Pro-Poor Economic Growth and Environmental Sustainable Development Poverty and Environmental Initiative (PEI): Support LGAS Working Sessions to Reflect P-E Objectives and Lessons Leant from Chololo Eco-Village. . Dar es salaam.

[22] de Jong, W. (2010). Forest rehabilitation and its implication for forest transition theory. Biotropica, *42*(1), 3-9.

[23] Prager, K., & Posthumus, H. (2010). Socio-economic factors influencing smallholder smallholder smallholder farmers' adoption of soil conservation practices in Europe. *Human Dimensions of Soil and Water Conservation". (Ed.): Napier, TL Nova Science Publishers, Last Accessed, 9*, 2014.

[24] URT. (2004). *The National Environmental Management Act*. Dar es Salaam, Tanzania: The Vice President's Office (Environment).

[25] Karki, L. B., & Bauer, S. (2004). *Technology adoption and household food security. Analyzing factors determining technology adoption and impact of project intervention: A case of smallholder peasants in Nepal.* Paper presented at the Proceedings of Deutscher Tropentag Workshop.

[26] Liang, W.-l., Carberry, P., Wang, G.-y., Lü, R.-h., Lü, H.-z., & Xia, A.-p. (2011). Quantifying the yield gap in wheat–maize cropping systems of the Hebei Plain, China. Field crops research, *124*(2), 180-185.

[27] Wang, H., Zhang, L., Dawes, W., & Liu, C. (2001). Improving water use efficiency of irrigated crops in the North China Plain—measurements and modelling. Agricultural water management, *48*(2), 151-167.

[28] Khapayi, M., & Celliers, P. (2016). Factors limiting and preventing emerging smallholder smallholder smallholder farmers to progress to commercial agricultural farming in the King William's Town area of the Eastern Cape Province, South Africa. South African Journal of Agricultural Extension, *44*(1), 25-41.

[29] Steduto, P., Hsiao, T. C., Fereres, E., & Raes, D. (2012). *Crop yield response to water* (Vol. 1028): fao Rome.

[30] Adesina, A. A., & Baidu-Forson, J. (1995). Farmers' perceptions and adoption of new agricultural technology: evidence from analysis in Burkina Faso and Guinea, West Africa. Agricultural economics, *13*(1), 1-9.

[31] Ball, J., Carle, J., & Del Lungo, A. (2005). Contribution of poplars and willows to sustainable forestry and rural development. UNASYLVA-FAO-, *56*(2), 3.

Chapter 10

Desertification in Agricultural Lands: Approaches to Mitigation

Mehdi Rastgoo and Alireza Hasanfard

Abstract

Urban expansion and industrial development destroy agricultural lands, pastures, and forests, and reduce the ecological and biological potential of lands, known as desertification. Diminished land potential due to one or a combination of processes such as erosion, destruction of water resources, destruction of vegetation, and swamping, by climate and human factors, is called desertification. Among these, human factors have a vital role in the emergence of this phenomenon. Excessive human economic activity upsets the ecological balance of arid and semi-arid regions, leading to adverse environmental changes. With the expansion of deserts in some parts of the world, food production and water resources are declining, resulting in environmental migration. Due to the limited capacity of urban areas to provide facilities and services, these migrations will cause severe socio-economic problems. In general, climatic and human factors are among the fundamental causes of desertification in the world. Preventing improper agricultural practices that lead to salinity and widespread soil degradation requires well-planned and strengthened awareness programs and development of information and care systems for areas exposed to desertification and drought, while also addressing the economic and social dimensions in these ecosystems.

Keywords: climate, desert, ecological balance, soil degradation

1. Introduction

Desertification is at the forefront of the environmental crises currently facing the international community. In sensitive and fragile desert-adapted ecosystems, degradation processes can easily be converted into an irreversible trend. Desertification reduces access to ecosystem services, increases food insecurity and poverty, and affects communities' well-being [1]. Desertification is land degradation or the impoverishment of arid, semi-arid (drylands), and some subhumid ecosystems, resulting from many factors including human activities and climatic change. The assessment of global scale desertification vulnerability to climate change and human activity is important to help decision-makers formulate the best strategies for land rehabilitation and combat global desertification in sensitive areas [2]. The range and intensity of desertification have increased in some dryland areas over the past several decades [3].

Drought and unreliable and variable rains are recurrent problems. Even without climate change, drylands face a daunting array of threats including population pressure, social changes (e.g. settlement of traditionally nomadic peoples), and exploitive agricultural and grazing practices that increase deforestation, soil erosion,

IntechOpen

salinization, and water depletion. Many political and institutional problems have conspired to degrade 20% of the world's drylands, including 22% of Asia's and 25% of Africa's susceptible drylands [4].

Regions like Africa are particularly vulnerable to desertification since two-thirds of the continent is made up of either deserts or drylands, while 73% of its agricultural drylands are already degraded. More than two-thirds of its population is composed of subsistence farmers, and, therefore, the impact of land degradation is immediate and devastating [5].

2. Desertification processes

Desertification may occur as a result of one process or the interaction of several functions. For example, wind erosion is one of the essential desertification processes in arid regions of the world, which alone or combined with other processes leads to desertification. The main desertification processes refer to the destruction of plant resources, soil resources, soil erosion, and water erosion, which are further explored below. In addition to natural factors, policies in Greece or Europe in recent decades have been reported to contribute to intensive land cultivation, overgrazing, rural–urban migration, etc., which directly affect desertification [6].

2.1 Destruction of plant resources

Degradation of vegetation through harvesting and destruction is the dominant desertification process. Ma et al. [7] cited vegetation degradation as an essential factor in southwest China's socio-economic development. The researchers also identified nutrients in the soil, especially N, P, and K, as the main factors influencing plant species composition in rocky desert areas. Vegetation conservation in Greece has been introduced as an influential factor in reducing water and wind erosion [6]. Vegetation cover and vegetation composition are the most common characteristics of many terrestrial ecosystems. These characteristics are associated with many ecosystem services, including biodiversity, soil and water conservation, food production, and fiber. It is also common to use these two indicators to assess land degradation and rehabilitation and rehabilitation project success. Deforestation contributes to about 17% of annual human greenhouse gas (GHG) emissions [8]. Humans resort to deforestation to meet their wood and energy needs. Deforestation for fuelwood is much more significant in developing countries with high populations and less access to commercial energy sources. Forests are also being destroyed to provide more land for agriculture [9]. Desertification risk scenarios in northeastern Brazil predict that 75% of forest areas will decline from 2010 to 2040. In this scenario, most forest areas will be replaced by agricultural lands [1]. Therefore, lands with more suitable vegetation are more resistant to degradation. In contrast, poor vegetation areas are fragile and accelerate desertification over time due to adverse environmental factors.

2.2 Destruction of soil resources

Land degradation in recent years has become a primary global concern due to increased waste disposal and demand for food production. Soil flexibility is limited, and soil degradation can never be easily reversed. In this century, the focus of land degradation has been on soil erosion, since forests, grasslands, and wetlands have been destroyed for crop production.

Severe land use without proper soil management, especially in fragile ecosystems, can accelerate desertification [10]. Human activities or climate change negatively affecting vegetation can lead to irreversible soil degradation in semi-arid regions [11]. Soil degradation in the semi-arid region of northeastern Brazil is driven by a limited set of variables, the most important of which are climatic, economic, and population growth variables. These factors lead to the expansion of agricultural lands and overgrazing, which increase the rate of deforestation [12].

Soil degradation processes:

2.2.1 Physical destruction

These refer to drastic changes in the soil's physical properties, including reduced permeability and porosity, reduced stability of the soil structure, and loosening and compaction of the soil [13]. Root zone compaction is the main form of physical degradation in arable lands and pastures, reducing soil fertility and reducing the amount of soil organic matter. Low structural stability of compacted soils leads to high vulnerability to mechanical stresses from agricultural operations. Therefore, reducing soil permeability, increasing runoff, increasing erosion, reducing soil aeration, and reducing biomass production are side effects of soil compaction and tuber formation that should be considered an indicator to assess the intensity of desertification.

2.2.2 Chemical degradation

A change in the soil's chemical properties in such a way that it interferes with nutrient uptake is called soil chemical degradation. Soil salinization, soil acidity imbalance, soil leaching, and ultimately reduced soil fertility are the most critical consequences of chemical soil degradation. Chemical degradation of soil can also occur due to increased concentrations of some toxic components such as aluminum.

2.2.3 Soil biodegradation

Microorganisms in agricultural soils play a crucial role in soil fertility. The reduction of soil organic matter and living microorganisms in soil is called biodegradation. Humus is an essential soil substance that increases soil porosity, soil stability, soil water holding capacity, and micronutrients. Organic matter depletion is the first state of biodegradation that leads to changes in other soil properties. In arid regions, depletion of soil organic matter leads to a decrease in soil moisture-holding capacity, a reduction in crops, and an increase in soil erosion [14]. Land-use change affects the physical, chemical, and biological properties of the soil. The conversion of pastures into agricultural lands in some areas of Iran has reduced the soil quality and increased soil degradation. It has been reported that the transformation of ranges to agricultural fields in three regions in Isfahan province has reduced soil organic matter by about 26% in agricultural lands, which is probably due to poor vegetation density (**Figure 1**) [15].

2.3 Soil and water erosion

In the last century, the significant destruction of land has been through soil erosion, as the areas of forests, grasslands, and wetlands have been removed for crop production. Soil erosion is one of the essential desertification processes during which soil particles are separated, transported, and deposited. Moreover, the soil decays and its organic matter decreases in the process of erosion. Humans obtain more than 99.7% of their food (calories) from land and less than 0.3% from the

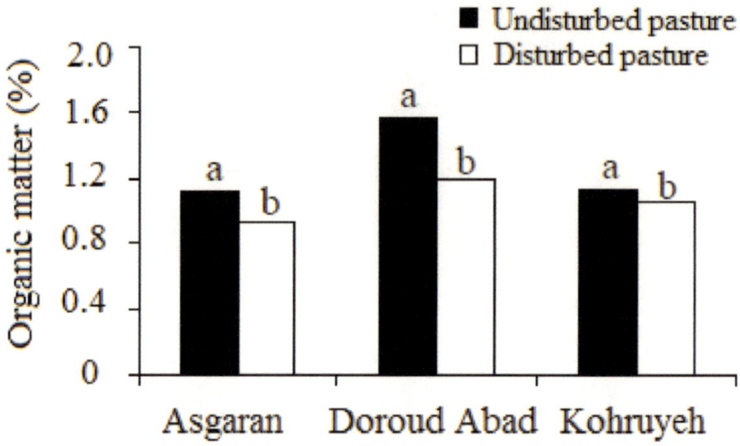

Figure 1.
Percentage of soil organic matter in the lands of three regions. The same letters for the regions indicate no statistical difference at the 5% level with the LSD test [15].

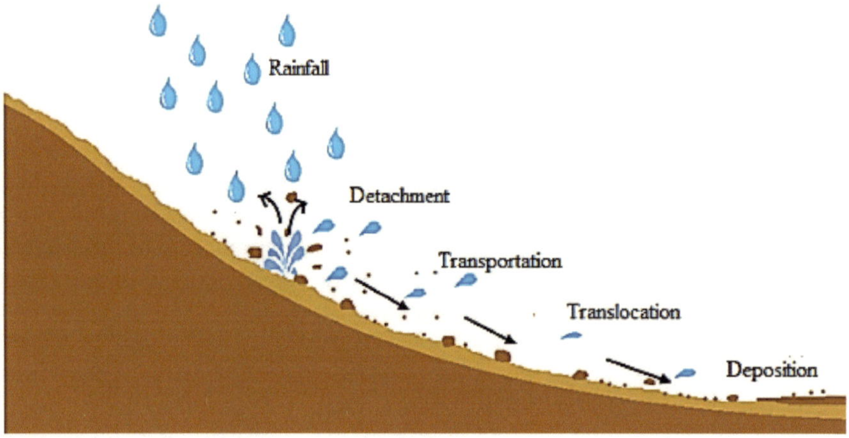

Figure 2.
Erosion stages by water [17].

oceans and other aquatic ecosystems. About 10 Mha of crops are lost each year due to soil erosion, thus reducing the amount of arable land available for food production [16]. Water erosion means the removal, transport, and deposition of soil particles by rain, runoff, and gravity **Figure 2** [18] shows the mechanism of water erosion. Rain erosion is one of the most critical water erosion types, which occurs more widely than other types. As rainfall occurs, the raindrops onto the soil surface and makes the first contact with the soil [19]. The loosening of soil particles or the detachment process takes place when soil particles disengage as the rain touches down on the soil. Afterward, soil particles are transported by rolling, splashing, or dragging and translocate to another place. Finally, soil particles are deposited at some other place at a lower elevation [17].

3. Factors affecting desertification

The process of land degradation in arid areas is called desertification, which affects the land capacity to provide ecosystem services, such as food production or biodiversity hosting, to name a few. It is stimulated by both human activity and climate and depends on the specific context. More than 1 billion people in about 100 countries face some of the risks associated with the effects of desertification [20]. The risk of desertification can increase in parts of the world that may become arid due to climate change.

According to the United Nations (UN), about 2.1 billion people worldwide (25–30%) living in arid and semi-arid regions [21] are among the poorest people. Approximately 70% of rainfed areas are located in Africa and Asia (**Figure 3**) [3]. Asia, followed by Africa and South America, have the largest populations in arid regions [23]. The global population growth rates are very high, especially in dry areas. The rapid population growth increases the pressure on land and natural resources that have already been oppressed and leads to poverty by degrading land and increasing desertification [24].

Parivar et al. [25] comparisons of open, green, and impervious surface areas (ha) in Yazd, Iran for 1991 and 2018 (**Figure 4**). The continuum, impermeable levels (built area) increased strongly during the period under study. From 1991 to 2018, there was an 80% decrease in open space, 63% decrease in green space, and a 90% increase in built-up area. In this way, population growth and urban development reduce green space, which leads to desertification.

Numerous factors affect the intensification and advancement of desertification, which can be classified into two categories, anthropogenic and climatic factors.

3.1 Anthropogenic factors

Human factors play an intensifying role in the development of desertification in arid and semi-arid ecosystems. These factors have a significant contribution to the process of land degradation. The annual plants are destroyed and the soil dries out when rainfall is low, thereby providing water and wind erosion and forming deserts and desertification globally. Some researchers have identified humans as a significant cause of desertification [4]. The shares of human factors and natural elements involved in desertification were 87 and 13%, respectively [26]. Changes in the use and destruction of forests and pastures, overgrazing, salinization of water and soil resources, burning of crop residues, improper use of groundwater, irregular

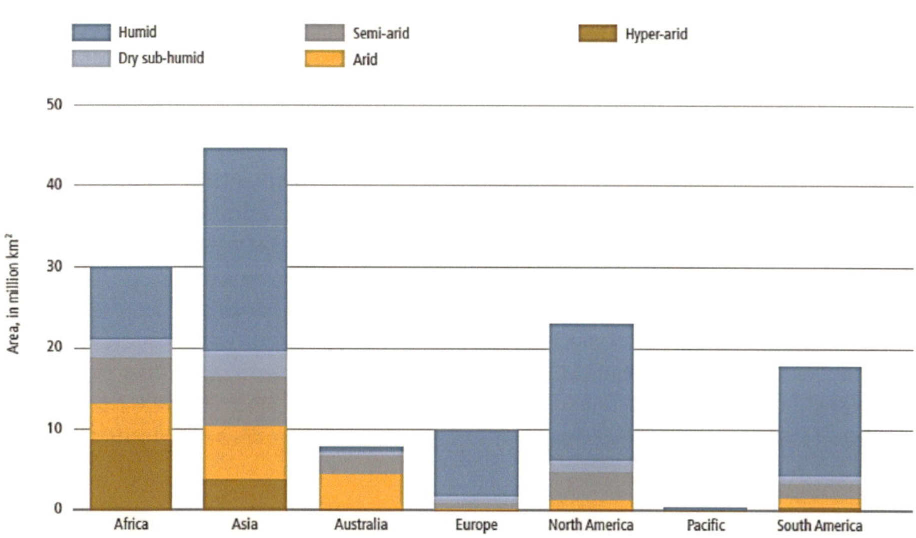

Figure 3.
Dryland categories across geographical areas (continents and Pacific region). Data: TerraClimate precipitation and potential evapotranspiration (1980–2015) [22].

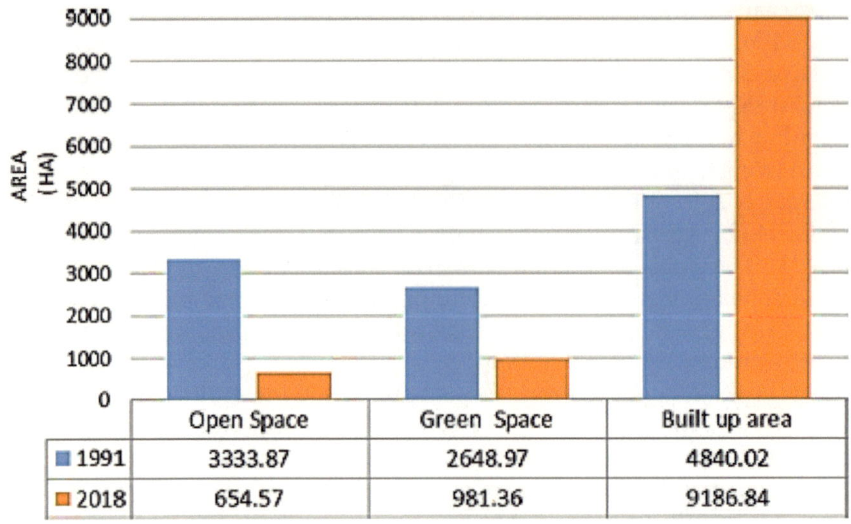

Figure 4.
Land areas (ha) in Yazd categorized as open, green, and built-up space (1991–2018) [25].

plowing, and failure to observe proper crop rotation in agriculture, fire, sediment settlement, development of urban areas, and industrial activities are among the most important factors influencing desertification by humans. Deserts are divided into two general categories—natural deserts of environmental origin and human deserts. Natural deserts are commonly found in arid and hyper-arid regions of the world, and humans have a minor role in their formation, making bare natural landscapes without vegetation. In addition to arid and hyper-arid regions, human deserts can be seen in other climatic zones, including semi-arid to humid regions.

3.2 Climatic factors

Climate change can further increase the risk of desertification for those regions of the world that may change into drylands for climatic reasons. Because arid areas are used for various purposes, such as agriculture, grazing, and gathering wood for fuel, the multiple activities performed in them can exacerbate the problem of desertification and bring about lasting changes in rainfed ecosystems. In this regard, drought, irregularity in rainfall, topographic factors, and radiation angle are the most critical environmental factors affecting desertification.

Although the effects of climate change can be seen in all activities, its impact on agricultural production seems to be more pronounced; annual damage from the adverse impacts of climate change due to rising temperatures, long periods of drought and desertification has been reported to be far greater than other activities affected by this change. For example, climate change has directly reduced crop productivity by reducing crop yields [27]. Barren ecosystems have low and variable rainfall, so climate change and other factors that lead to prolonged drought can rapidly reduce these ecosystems biological productivity. These changes may be temporary and last only one season, or they may last for years and decades [3].

4. The role of agriculture in desertification

One of the most important causes of desertification, especially in arid and semi-arid regions of the world, is improper activities in agriculture. Dense and improper cultivation of crops reduce soil structure stability and lead to soil degradation and erosion [28]. When the soil's pressure increases due to agricultural operations and land clearing, soil fertility decreases, resulting in soil degradation and erosion.

Irrigation systems have developed over time as an agricultural technique in arid areas with low rainfall. If irrigation methods are misapplied, water loss and lowering of groundwater aquifers will lead to salinization and alkalinization of lands. Therefore, one of the most critical factors in desertification in agriculture is improper irrigation, which will ultimately lead to the destructive effects of drought, the phenomenon of soil and water salinization due to improper management of agricultural land [28]. Salinity is caused by improper irrigation in soils of arid and semi-arid regions. The leading cause of salinity and alkalinity in some parts of the world is the entry of low-quality water. Improper irrigation and lack of proper drainage raise the groundwater level and form a surface saline aquifer. With soil degradation, vegetation decreases and the soil is exposed to water and wind erosion, and its fertility is severely reduced. Therefore, knowing the main reasons for the salinization of soil and water in the world's regions and proper management of such lands can help reduce the process of desertification.

Changes in the soil's chemical composition with the use of fertilizers or chemical pesticides lead to changes in the soil's physical condition, which increases soil erosion. On the other hand, nitrogen released from chemical fertilizers and its mixing with groundwater lead to pollution of groundwater resources, which is recognized as a severe problem in some parts of the world [29].

Effects of fire on vegetation cover change and expedite the process of desertification by humans is proven. The severity of the fire damage depends on the conditions of the area. In arid lands, severe fires lead to the extinction of plant, animal, and soil species, which in some cases alter ecosystems and contribute to intensified desertification. Frequent burning of straw and crop residues and reducing land fertility are also influential factors in desertification.

There is an important concern about the conversion of agricultural land to barren desert, exacerbated by desertification in countries such as Iran.

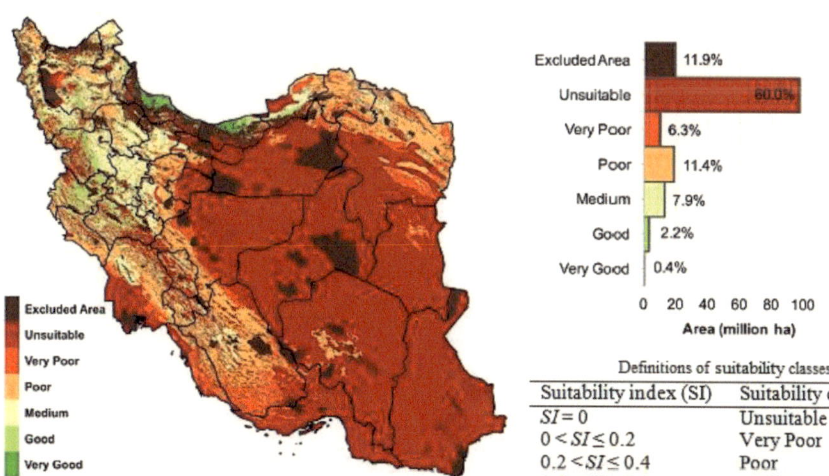

Definitions of suitability classes

Suitability index (SI)	Suitability class (SC)
$SI = 0$	Unsuitable
$0 < SI \leq 0.2$	Very Poor
$0.2 < SI \leq 0.4$	Poor
$0.4 < SI \leq 0.6$	Medium
$0.6 < SI \leq 0.8$	Good
$SI > 0.8$	Very Good

Figure 5.
Classification of Iran's land suitability for agriculture based on soil properties, terrain, and climate conditions [30].

Accordingly, desertification is one of the most important factors in Iran, which has been exacerbated by drought. The result of drought and desertification is a reduction in Iran's land suitable for agriculture. Mesgaran et al. [30] classified Iran's land suitability for cropping as (million ha): very good 0.4% (0.6), good 2.2% (3.6), medium 7.9% (12.8), poor 11.4% (18.5), very poor 6.3% (10.2), unsuitable 60.0% (97.4), and excluded areas 11.9% (19.3) (**Figure 5**). Hence, reducing suitable land for agriculture means reducing food production, which threatens food security.

5. Desertification and food security

The most critical role of the agricultural sector is to ensure sustainable food security. So food security is, by definition, a situation where everyone has access to adequate, healthy, and nutritious food [31]. Therefore, desertification can be considered one of the most important factors limiting agricultural product production and, ultimately, the challenge of food security. According to **Figure 6**, if the population grows at a fixed exponential rate, the amount of food required will increase exponentially. But Malthus held that the output of food could increase only by a constant amount each period. Given these two different growth processes, food requirements would eventually catch up with food production. The population hits the subsistence level of food production at the Malthusian trap, shown here at point T [32].

Drought stress in arid and semi-arid regions such as Iran has posed a serious challenge to sustainable production to provide food for the growing population. Concerns about the vulnerability of agricultural production become more pronounced when there is a proper understanding of the impact of climate change. If desertification leads to degradation and degradation of water, soil, and vegetation resources as three factors of survival in vulnerable ecosystems, food security will face serious problems. Thelma [33] reported that desertification has exacerbated the problem of food security in eleven states in northern Nigeria and its effect is very glaring on the agricultural sector.

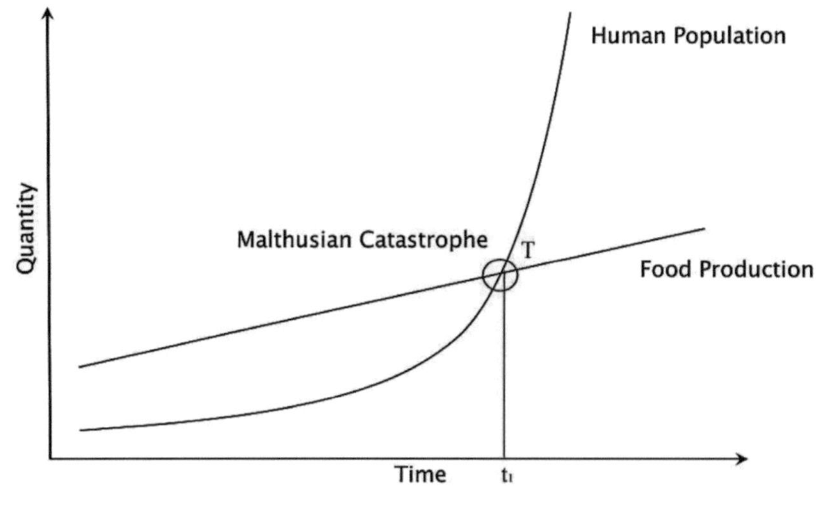

Figure 6.
The Malthusian growth model [32].

6. Strategies to combat desertification

In 1977, at the United Nations Conference on Desertification (UNCOD) in Nairobi, Kenya, representatives, and delegates first contemplated desertification's worldwide effects. The conference explored the causes and contributing factors and also possible local and regional solutions to the phenomenon. Also, the delegates considered the varied consequences of desertification, such as crop failures or decreased yields in rain-fed farmland, the loss of perennial plant cover and thus loss of forage for livestock, reduced woody biomass, and therefore scarcity of fuelwood and building materials, a decrease in potable water stocks from reductions in surface water and groundwater flow, increased dune intrusion onto croplands and settlements, increased flooding due to rising sedimentation in rivers and lakes, and amplified air and water pollution from dust and sedimentation [34].

Non-desertification means preventing desertification of areas exposed to desertification due to destructive human actions, not destroying existing natural deserts. Climate, as the two main climatic factors that have many changes and fluctuations, plays an essential role in forming arid and desert areas. Since arid and semi-arid regions occupy a large part of the world area, identifying the potentials in these areas and discovering resources to provide management solutions to take advantage of existing potentials and prevent desertification from Is of particular importance [3].

The priority in combating this phenomenon is to prevent the destruction of lands that have not yet been destroyed or where the descent rate is low. In contrast, for the conquered lands, remedial plans should be considered. According to international experts, including the following in the medium and long-term goals can significantly improve desert management activities.

- Strengthen awareness and development of information and care systems for areas exposed to desertification and drought, including the economic and social dimensions of these ecosystems

- Development of comprehensive desertification plans and their inclusion in national development plans and programs and national environmental plans.

- Combating land degradation through soil conservation, afforestation, and reforestation.

- Develop comprehensive compensation and mobilization programs to combat drought, including self-help arrangements for drought-prone areas and programs for environmental refugees.

- Develop coordinated development programs to eradicate poverty and improve living standards in areas prone to desertification.

- Encourage and improve the level of public cooperation and environmental education with a focus on desertification control and drought management

7. Conclusion

Because rainfed environments are used for a variety of human purposes (such as agriculture, livestock grazing, and timber collection), the various activities performed in them can exacerbate desertification problems and bring about lasting changes in rainfed ecosystems. Desertification is a phenomenon that occurs as a result of natural or human factors. In recent decades, Desertification has intensified with the loss of fertile lands, destruction of pastures and agricultural lands, salinization and erosion of lands, and quantitative and qualitative reduction of surface and groundwater. This phenomenon poses a serious threat to most of the world's ecosystems, mainly arid and semi-arid regions.

In recent years, due to the overexploitation of natural resources, this process has become more intense and severe, and comprehensive management is needed to deal with it. Implementing and implementing projects appropriate to this challenge and benefiting from successful practical programs in the world can effectively reduce the desertification process, especially in arid and semi-arid regions. In general, correct knowledge and assessment of the state of the world's deserts will help manage these areas.

Author details

Mehdi Rastgoo* and Alireza Hasanfard
Department of Agrotechnology, Faculty of Agriculture, Ferdowsi University of Mashhad, Iran

*Address all correspondence to: m.rastgoo@um.ac.ir

References

[1] Vieira, R. M. D., Tomasella, J., Barbosa, A. A., Martins, M. A., Rodriguez, D. A., Rezende, F. S., Carriello, F & Santana, M. D. (2021). Desertification risk assessment in Northeast Brazil: Current trends and future scenarios. Land Degradation & Development, 32(1), 224-240.

[2] Huang, J., Zhang, G., Zhang, Y., Guan, X., Wei, Y., & Guo, R. (2020). Global desertification vulnerability to climate change and human activities. Land Degradation & Development, 31(11), 1380-1391.

[3] Mirzabaev, A., Wu, J., Evans, J., Garcia-Oliva, F., Hussein, I. A. G., Iqbal, M. M., et al. (2019). Desertification. In Climate change and land. An IPCC Special Report on climate change, desertification, land degradation, sustainable land management, food

[4] Abdi, O. A., Glover, E. K., & Luukkanen, O. (2013). Causes and impacts of land degradation and desertification: Case study of the Sudan. International Journal of Agriculture and Forestry, 3(2), 40-51.

[5] Penny, R. (2009). Desertification and deforestation in Africa. Land use, land cover and soil sciences, 5, 201-219.

[6] Yassoglou, N., Tsadilas, C., & Kosmas, C. (2017). Land degradation and desertification. In The Soils of Greece (pp. 87-96). Springer, Cham.

[7] Ma, T., Deng, X., Chen, L., & Xiang, W. (2020). The soil properties and their effects on plant diversity in different degrees of rocky desertification. Science of The Total Environment, 736, 139667.

[8] Gorte, R. W., & Sheikh, P. A. (2010). Deforestation and climate change (CRS Report for Congress. Congressional ResearchService). Retrieved fromwww.crs.gov

[9] Katyal, J. C., & Vlek, P. L. (2000). Desertification: concept, causes and amelioration (No. 33). ZEF Discussion papers on development policy.

[10] Vieira, R.M., Tomasella, J., Alvalá, R.C., Sestini, M.F., Affonso, A.G., Rodriguez, D.A., Barbosa, A.A., Cunha, A.P., Valles, G.F., Crepani, E., De Oliveira, S.B., De Souza, M.S., Calil, P.M., De Carvalho, M.A., Valeriano, D.M., Campello, F.C., Santana, M.O., 2015.Identifying areas susceptible to desertification in the Brazilian northeast. Solid Earth, 6, 347-360.

[11] Albaladejo, J., Martinez-Mena, M., Roldan, A., & Castillo, V. (1998). Soil degradation and desertification induced by vegetation removal in a semiarid environment. Soil use and Management, 14(1), 1-5.

[12] Geist, H. J., & Lambin, E. F. (2004). Dynamic causal patterns of desertification. Bioscience, 54(9), 817-829.

[13] Poesen, J. W., & Nearing, M. A. (1993). Soil surface sealing and crusting. Proceedings, Catena Supplement 24 (1993): 139-140.

[14] Werner, W. J., Sanderman, J., & Melillo, J. M. (2020). Decreased soil organic matter in a long-term soil warming experiment lowers soil water holding capacity and affects soil thermal and hydrological buffering. Journal of Geophysical Research: Biogeosciences, 125(4), e2019JG005158.

[15] Hajabbasi M, Besalatpour A, Melali A. (2008). Impacts of Converting Rangelands to Cultivated Land on Physical and Chemical Properties of Soils in West and Southwest of Isfahan. Journal of Hydrology and Soil Science. 11 (42) :525-534.

[16] Pimentel, D. (2006). Soil erosion: a food and environmental threat.

Environment, development and sustainability, 8(1), 119-137.

[17] Soil Web. SoilWeb/Soil Management/Soil Erosion; The University of British Columbia: 2003. Available online: https://wiki.ubc.ca/LFS:SoilWeb/Soil_Management/Soil_Erosion (accessed on 25 March 2021).

[18] Foster, G. R., Young, R. A., Römkens, M. J. M., & Onstad, C. A. (1985). Processes of soil erosion by water. Soil erosion and crop productivity, 137-162.

[19] Zafirah, N., Nurin, N. A., Samsurijan, M. S., Zuknik, M. H., Rafatullah, M., & Syakir, M. I. (2017). Sustainable ecosystem services framework for tropical catchment management: A review. Sustainability, 9(4), 546.

[20] Koné, A., Fontaine, A., & El Yacoubi, S. (2020). Coupling cellular automata with medalus assessment for the desertification issue. In book of abstracts (p. 89).

[21] United Nations. (2019). (UN) Retrieved from https://www.un.org/en/events/desertific ation decad e/whyno w.shtml. Accessed 16 May 2019.

[22] Abatzoglou, J. T., Dobrowski, S. Z., Parks, S. A., & Hegewisch, K. C. (2018). TerraClimate, a high-resolution global dataset of monthly climate and climatic water balance from 1958-2015. Scientific data, 5, 1-12.

[23] FAO. 2008. Cereals and drylands. By Koohafkan, P. and Stewart, B. A. The Food and Agriculture Organization of the United Nations and Earthscan. Rome, FAO.

[24] Veron, S. R., Paruelo, J. M., & Oesterheld, M. (2006). Assessing desertification. Journal of Arid Environments, 66, 751-763.

[25] Parivar, P., Quanrud, D., Sotoudeh, A., & Abolhasani, M. (2020). Evaluation of urban ecological sustainability in arid lands (case study: Yazd-Iran). Environment, Development and Sustainability, 1-30.

[26] Babaev, A. G. (1985). Methodological principals of desertification processes assessment and mapping. *Turkmenistan: Desert Research Institute. Turkmenistan*.

[27] Korres, N.E.; Norsworthy, J.K.; Tehranchian, P.; Gitsopoulos, T.K.; Loka, D.A.; Oosterhuis, D. M.; Gealy, D.R.; Moss, S.R.; Burgos, N.R.; Miller, M.R and Palhano, M. (2016). "Cultivars to face climate change effects on crops and weeds: a review". Agronomy for Sustainable Development. 36(1): 12.

[28] Danfeng, S., Dawson, R., & Baoguo, L. (2006). Agricultural causes of desertification risk in Minqin, China. Journal of environmental management, 79, 348-356.

[29] Norton, D., Shainberg, I., Cihacek, L., & Edwards, J. H. (1999). Erosion and soil chemical properties. *Soil Quality and Soil Erosion, Soil and Water Conservation Society, Ankeny*.

[30] Mesgaran, M. B., Madani, K., Hashemi, H., & Azadi, P. (2017). Iran's land suitability for agriculture. Scientific reports, 7(1), 1-12.

[31] Pinstrup-Andersen, P. (2009). Food security: definition and measurement. Food security, 1, 5-7.

[32] Coale, A. J., & Hoover, E. M. (2015). *Population growth and economic development*. Princeton University Press.

[33] Thelma, M. N. (2015). Desertification in northern Nigeria: Causes and implications for national food security. *Peak Journal of social sciences and* Humanities, 3, 22-31.

[34] United Nations Conference on Desertification [Internet] 1977. Available from: https://www.britannica.com/topic/United-Nations-Conference-on-Desertification [Accessed: April 23, 2021]

Chapter 11

Combating Desertification through Enhancement of Woody Floral Diversity in the Drylands of Kenya: Analysis, Milestones, and Strategies

Joseph Hitimana, Edward K. Mengich, Teresiah N. Kuria and Pauline Kimani

Abstract

Desertification remains one of the most challenging phenomena in the drylands of Kenya, where it affects about 80% of the country. This is because of persistent degradation of these areas by climatic variations, human activities, and overgrazing by livestock and wildlife. In these areas, inhabitants suffer from widespread acute poverty and other adverse effects of drought. In order to effectively and efficiently combat desertification and reduce the impacts of further degradation, the Government of Kenya and partners are committed to developing and implementing methods, approaches, strategies, and mechanisms that would slow down or reverse this phenomenon. This chapter covers an in-depth review of advances made so far in the area of woody resources restoration and sustainable management in the drylands of Kenya through biodiversity assessments, conservation, rehabilitation, afforestation, and reafforestation initiatives and research. Achievements, challenges, and opportunities encountered are highlighted for sustainable development and wise utilization of dryland woody and allied resources.

Keywords: degradation, desertification, drylands, floral diversity, natural resources, species adaptability

1. Introduction

Desertification is one of the most important challenges to livelihoods and development in the drylands [1, 2]. Traditionally, the drylands of Kenya are vast and rich in biodiversity and natural resources [3]. However, recent decades have seen increased human pressure on forests and woodlands that has created conditions conducive to degradation, deforestation, and desertification, thus reducing national tree cover to all time unacceptable levels. Deforestation leads to the deterioration of soil fertility, which occurs rapidly under tropical climates. However, it also offers a great potential for intensified afforestation toward achieving the national objective of 10% tree cover [2, 4]. This chapter provides an in-depth analysis of the problem including the extent, genesis, impacts, and remedies instituted by the Government of Kenya, her citizens, and development partners. The authors present an in-depth review of relevant reports and case studies and share original data and maps in view to present a favorable case toward promoting greening of drylands in Kenya.

2. Desertification process and Kenya's context

Drylands occupy 41% of the earth's land surface and are home to 35% of its population. They occur in every continent but are most extensive in Africa. The drylands include desert, grassland, and savanna woodland biomes. In Kenya, the drylands make up 84% of its total land surface, support about 9.9 million Kenyans (about 34% of the country's population), and account for more than 80% of the country's ecotourism interests, 60% of the nation's livestock, and up to 75% of the national wildlife population [5, 6]. Although rich in natural resources, the increased human pressure on forests and woodlands has created conditions conducive to degradation, deforestation, and desertification. The drylands environment poses formidable problems for sustainable development. Among these are unpredictable and severe drought, desiccation or aridification due to persistent drought, and dryland degradation or desertification [7]. However, drylands in Kenya are vast and offer a great potential for intensified afforestation toward achieving the national objective of 10% tree cover.

Desertification is defined as land degradation in arid, semiarid, and dry subhumid areas resulting from many factors, including climatic variations and human activities. These areas are characterized by low and erratic rainfall, high evapotranspiration, shallow soils with low water-holding capacity, and low soil fertility [8]. Drought is a common occurrence in the Arid and Semi-Arid Lands (ASALs) and is exacerbated by climate change [9]. It is caused by rainfall deficit; it leads to shortage of water and unusually high temperatures. Anthropogenic causes of desertification include overgrazing, deforestation, and removal of the natural vegetation cover by taking too much fuelwood, the build-up of salt in irrigated soils, topsoil erosion, and agricultural activities in the vulnerable ecosystems of arid and semiarid areas that are thus strained beyond their capacity. These activities are triggered by population growth, the impact of the market economy, and poverty. The phenomenon reduces agricultural output, contributes to droughts, and increases human vulnerability to climate change.

The differences and interlinkages between desertification, drought, desiccation, and climate change and their causal factors have been outlined in many texts [7]. Desertification is a type of land degradation in drylands in which biological productivity is lost due to natural processes or induced by human activities whereby fertile areas become increasingly arid [10]. Land degradation is a process in which the value of the biophysical environment is affected by a combination of human-induced processes acting upon the land [11]. It is viewed as any change or disturbance to the land perceived to be deleterious or undesirable. Permanent changes in climate, particularly rainfall, are responsible for natural desertification. Desertification may alter the living conditions of the local flora and fauna that makes it impossible for animals and plants to sustain their populations. After desertification, regions suffer from water shortages due to climate change and animals may suffer and die since water is vital for all life on the planet. Desertification results in persistent degradation of dryland and fragile ecosystems due to man-made activities and variations in climate. Desertification, in short, is when land that was of another type of biome turns into a desert biome because of changes of all sorts.

Desertification affects topsoil, groundwater reserves, surface runoff, human, animal, and plant populations. A study conducted in the Mutomo District, Kenya, confirmed that the main use of selectively harvested trees was charcoal production [12]. This consequently led to degradation of the woodlands through reduction in tree species richness, diversity, and density. Water scarcity in drylands limits the production of wood, crops, forage, and other services that ecosystems provide to our local communities. The United Nations Convention to Combat Desertification states on its website that, globally, more than 12 million hectares of land are lost annually to desertification, drought, and degradation and that over 1.5 billion people are directly dependent on land that is being degraded, leading to loss of US dollars-equivalent billions of earnings each year [13]. In Africa, three million hectares of forest along with an estimated 3% of Gross Domestic Product (GDP) are lost annually due to depleted soils. The result is that two-thirds of Africa's forests, farmlands, and pastures are now degraded.

The dry lands on average receive an annual rainfall of between 250 and 1000 mm. **Figure 1** shows the extent and levels of aridity in Kenya. The rains are typically of short duration but of high intensity and therefore highly erosive. The rate of evapotranspiration is also high. The aridity values on the map legend are based on the generalized climate classification scheme for Aridity Index values [15] as follows (**Table 1**). No region in Kenya is classified as Hyperarid (Aridity Index Value <0.03).

The main challenge in developing dry lands is how to increase availability and access to information and technology for the development and management of natural resources.

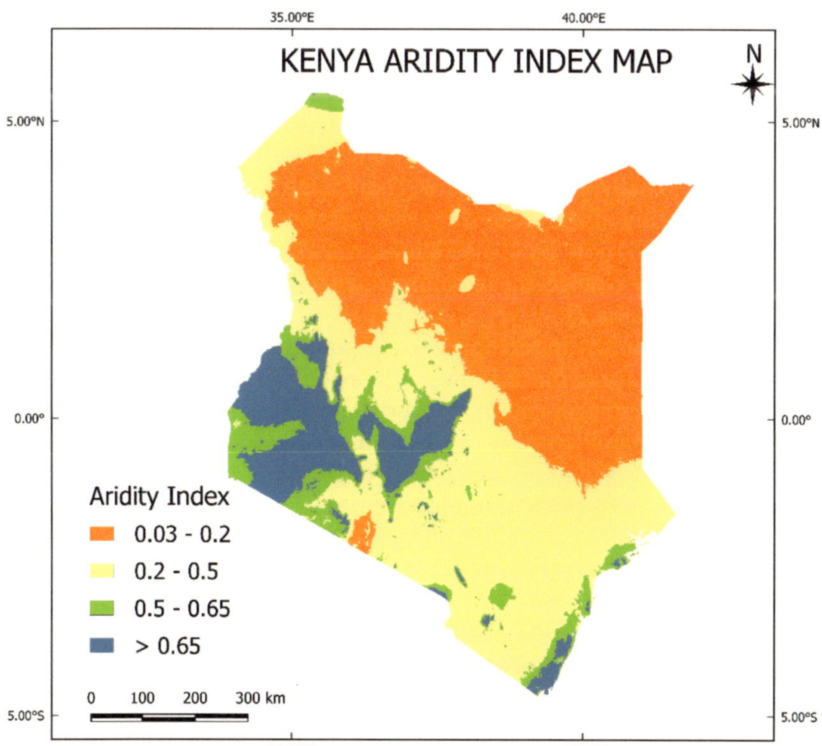

Figure 1.
Kenya aridity index map produced based on the Global Aridity Index and Potential Evapotranspiration (ET0) Climate Database v2 [14].

Aridity Index Value	Climate	Extent coverage (km^2)
0.03–0.2	Arid	255,800.11
0.2–0.5	Semiarid	220,704.78
0.5–0.65	Dry subhumid	46,483.89
>0.65	Humid	57,378.22

Table 1.
Aridity context in Kenya.

3. Land use, cover, and natural water resources

Overgrazing is the major cause of desertification in the dry lands [16]. Other factors that cause desertification include urbanization, climate change, overuse of ground water, deforestation, natural disasters, and tillage practices in agriculture that make soils more vulnerable to wind [16, 17].

3.1 Contextualized land uses in Kenya

A land-use map represents the physical and biological cover over the land surface, consisting of vegetation, bare soil, water, and artificial structures. Land-use and land-cover information is significant in understanding the socioeconomic and environmental implications linked to the utilization of the available natural resources in a region. In Kenya, land use is classified into agriculture, forest, bushland, grassland, plantation, built-up and urban area, barren land, woodland, plantation, swamp, and water bodies (**Figure 2**).

Agriculture is estimated to occupy 48% of the total land area, out of which 9.8% is considered arable land, 37.4% is covered by permanent pasture, and 0.9% by permanent crops such as tea and coffee plantations. Tree cover is estimated at 6% while other land uses, such as urban areas and bare land, occupy about 45.8% [19]. Urbanization and expansion of agricultural land have increased the rate of conversion and fragmentation of the natural forest ecosystems leading to deforestation and eventually land degradation. Land-use/land-cover change is considered the primary causal agent of climate change due to environmental changes that lead to increased greenhouse gas emissions. However, the effects of climate change, viz. increased temperature and variability in precipitation, prompt the change in land use as communities try to adapt to the changing climate [20]. Therefore, the development of effective land-use management plans is crucial to ensuring Kenya's goals toward environmental sustainability under future climate scenarios. There is a need to assess the vulnerability and adaptive capacity of local communities so as to prioritize the solutions that will enable communities to cope and adapt to climate change.

3.2 Contextualized tree cover types in Kenya

Kenya's forest areas constitute a wide range of vegetation, viz. trees, shrubs, and grass species. The Kenya Tree Cover Types map (**Figure 3**) gives a detailed visual representation of the categories of different tree cover types found in Kenya. The tree cover types are mainly classified as open canopy, closed canopy, multilayered trees, and mangrove trees. Open canopy refers to a collection of relatively tall trees that are spaced and allow easy penetration of sunlight to the ground surface. Closed canopy forest is a thicket of mature trees whose leaves and branches are densely spaced

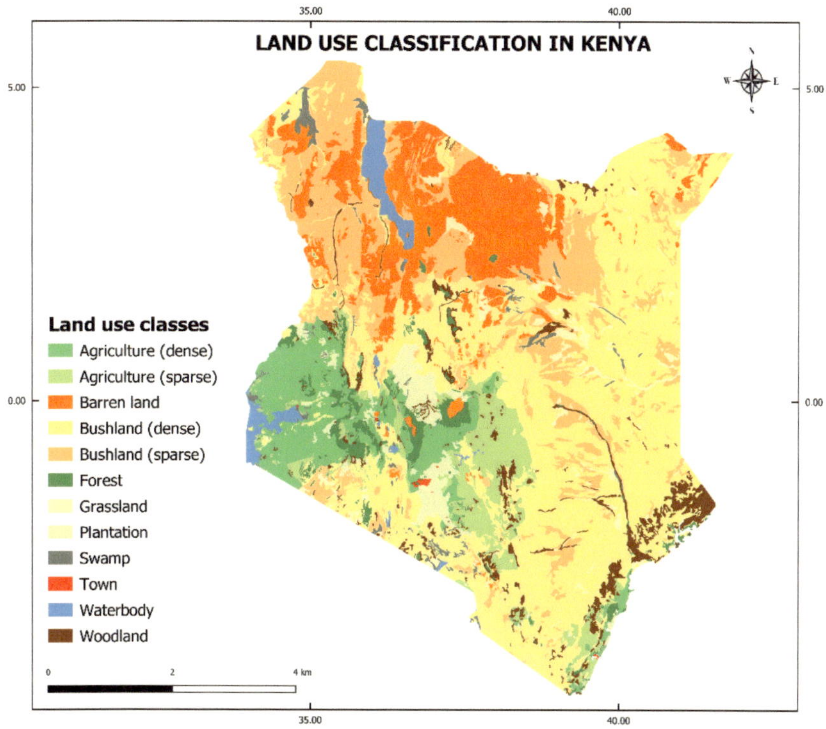

Figure 2.
Mapped land-use classification for Kenya based on Kenya's geospatial data provided by the World Resources Institute [18].

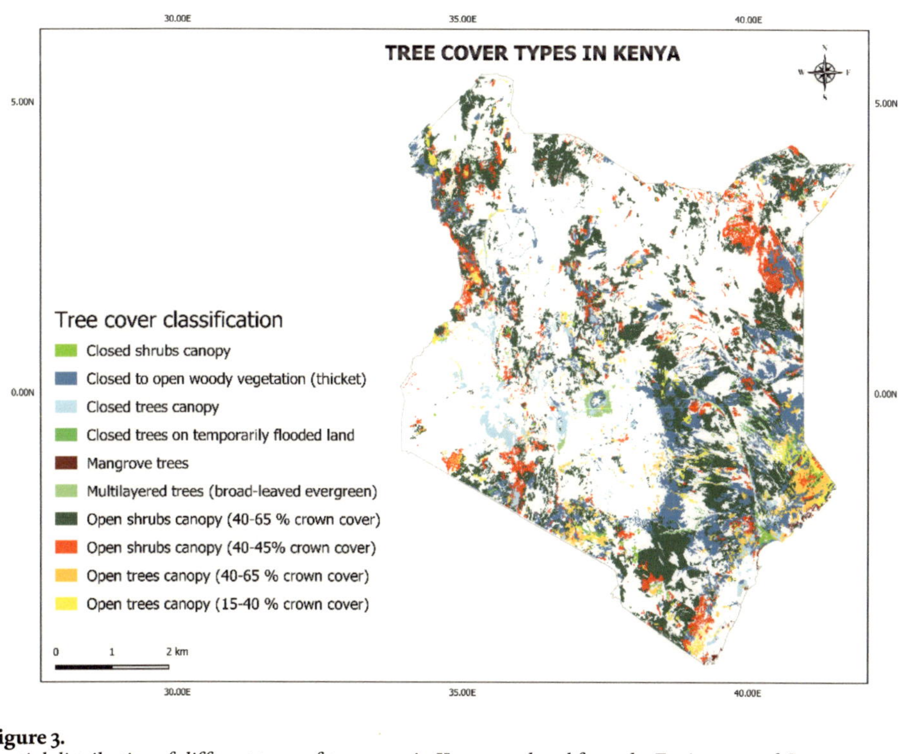

Figure 3.
Spatial distribution of different types of tree cover in Kenya produced from the Environmental Systems Research Institute Kenya GIS data [21].

creating a crown that encloses the understory and the forest floor. The canopy types are also distinguished based on the percentage of crown cover, that is, the proportion of the ground covered by the vertical projection of the tree canopy. The multilayered trees generally refer to the dense tropical evergreen forests having the appearance of structured layers that differ in the amount of sunlight penetration, ground cover, and water availability. The mangrove trees, which occupy about 1% of the land area in Kenya, include thickets along the coastlines, tidal estuaries, and salt marshes of Kenya. *Rhizophora mucronata* Lam is the principal species in most sites dominated by mangroves along the Kenyan Coast [22].

The closed canopy and multilayered trees each cover about 2% of Kenya's total land area and are mainly restricted to areas below an altitude of 3000 meters. Tree species dominance within this area is according to the Agro-Ecological Zone (AEZ) [23]. Moist forests occurring at 2100–3300 m above sea level, with rainfall above 1500 mm, are dominated by a variety of broad-leaved species that include *Tabernaemontana stapfiana* (Britten), *Dombeya goetzenii* (K. Schum), *Dracaena afromontana* (Mildbr), *Hagenia abyssinica* (Bruce J. F. Gmel), *Nuxia congesta* (Fresen), *Croton macrostachyus* (Delile), and *Podocarpus latifolius* (Thunb. Mirb) [23]. The drier montane forests occurring at 1800–2900 m a.s.l. with an annual rainfall of 700–1350 mm are characterized by species such as *Juniperus procera* (Endl), *Olea europaea ssp. africana* (Mill. P. Green), *Podocarpus falcatus* (Mirb.), *Cassipourea malosana* (Baker. Alston), *Acokanthera schimperi* (A. DC. Schweinf), *Ekebergia capensis* (Sparrm.), *Olinia rochetiana* (A. Juss.), *Teclea nobilis* (Delile), *Croton megalocarpus* (Hutch), and *Calodendrum capense* (L. f. Thunb) [23].

The open canopy trees and shrubs, woodland, and grassland vegetation are estimated to cover 65% of the total land area and represent the dominant vegetation type in the Arid and Semi-Arid Lands of Kenya [24, 25], with the dominant tree species being *Acacia mellifera*, *Acacia senegal*, *Acacia reficiens*, *Acacia tortilis*, and *Commiphora* sp. [26]. Presently, climate change is a major driver of the rate of encroachment into forested areas and the destruction of grasslands in the country. Climate change has led to competition for arable land resources due to declining water resources coupled with droughts and scarcity of arable land.

Evidence of climate change in Kenya has been manifested in the increased frequency of droughts and floods, changes in rainfall intensity and distribution patterns, and increased minimum and maximum temperatures. Future climate projections estimate that most regions in Kenya will experience a 100-mm decrease in long-season (March–May) rainfall by 2025 [27]. Increased climatic variability poses a threat to the trees and associated vegetation cover with regions such as the ASALs being most vulnerable to the adverse impacts. The projected climatic impacts are likely to exacerbate the rate of land degradation and desertification in the country.

3.3 Contextualized water resources in Kenya

Figure 4 shows the distribution of Kenya water resources. The country relies mainly on freshwater resources represented by lakes, rivers, swamps, and springs as well as dams, water pans, and groundwater. Kenya's annual freshwater resources endowment is estimated to be 20.2 BCM (billion cubic meters) or 548 m^3 per capita per year [28]. These surface water resources are highly dependent on the country's forested areas and highland ecosystems that serve as water catchments. The watersheds depicted on the map include the Ewaso-Samburu, Mt. Kenya and Aberdares, Mau and Western, Amboseli and Chyulu, and the Coastal forest and Marine watersheds. The major lakes in the country include, inter alia, Lake Victoria, Lake Nakuru, Lake Naivasha, and Lake Elementaita. Some of the major rivers include the Tana River, Mara River, Athi-Galana-Sabaki River, Tsavo River, Ewaso Ng'iro River,

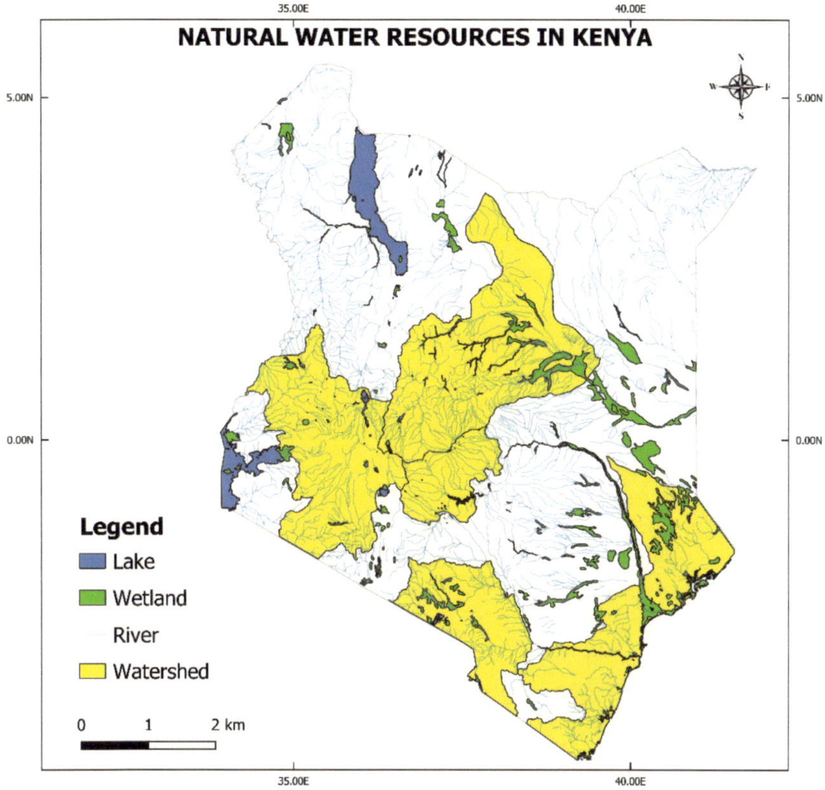

Figure 4.
Map of natural water resources in Kenya based on Kenya GIS data provided by the Environmental Systems Research Institute Kenya GIS data [26].

and Nzoia Rivers. The wetlands represented on the map are classified into fresh water marshes, saline/alkaline marshes, mangrove, lake shorelines, and saltpans [29]. Despite harboring some of the great water towers of East Africa, Kenya is among water-scarce countries, with its per capita renewable freshwater potential being 235 m^3 per annum [30].

Coupled with extreme climatic events, such as droughts and floods, the water resources in the country continue to show a declining trend and threaten the availability of water for economic and domestic use. In the past, prolonged dry seasons have led to 37% decline in the water levels in dams and other reservoirs subsequently causing crop failure, loss of livestock, and limited access to freshwater [31]. Additionally, increased frequency of floods has led to the destruction of land resources due to soil erosion, disruption of water supply systems, and the contamination of freshwater resources. Overall, climate change, destruction of water catchment areas and deforestation continue to increase water scarcity in Kenya.

4. Preventive and corrective initiatives

4.1 Background on combating desertification in Kenya

Like many other nations around the world, Kenya is threatened by desertification, land degradation, and drought. In some of the dryland areas such as in the North and North-Eastern Kenya, the deserts have eaten the once-potential landscapes turning them into inhabitable landscapes that cannot support humans, livestock, and even wildlife [3]. The government of Kenya, recognizing the importance of the country's dry lands to the country's socioeconomic development and realizing that they are being degraded fast, has sought ways of restoring and rehabilitating dry forests and woodlands, among other ecosystems in these lands. These ecosystems are comprised of trees that are specially adapted to the harsh climatic and edaphic conditions, providing important ecosystem services for communities in an environment where other types of tropical tree species would not survive. These trees are threatened and call for restoration and rehabilitation initiatives by the government through support and participation in thematic forestry research and development activities. These activities include restoration and sustainable management of woody resources through biodiversity assessments, conservation, rehabilitation, afforestation, and reafforestation in order to enhance plant diversity.

4.2 Role of enhancement of woody floral diversity in the drylands

Floral diversity refers to the diversity of plants occurring in a specific region during a particular era. It generally refers to the diversity of naturally occurring indigenous or native plants. The word "Flora" comes from the Latin word Flora, which means the goddess of plants. As Kenya has a limited area covered by indigenous timber-producing forests, plantations of exotic trees, mainly eucalypts, were established in the country in the early 1900s. These species were suitable to small-scale farmers and provided overall support to key sectors of the economy. The demand for timber had exceeded the supply available from indigenous forests; hence, the exotic species were preferred for afforestation because none of the indigenous tree species that yielded useful timber grew at rates considered profitable. Efforts to rehabilitate the dry lands are in place and include the promotion and establishment of suitable multipurpose tree species in the ASALs as well as water harvesting and conservation measures.

As rangelands, areas in ASALs have a relatively low production potential, are fragile, and are easily degradable through overutilization or use of inappropriate technologies [32]. There is a need to develop suitable ASAL rehabilitation technologies and to uphold efforts employed in providing solutions that sustainably improve the lands' productivity and combat desertification. Woody vegetation is one such renewable resource with an exceptional potential to provide the dry season's forage for livestock and serve as soil cover. Forests and woodlands are also biologically important because of the diverse fauna and flora associated with them. They, therefore, contribute significantly to the livelihoods and welfare of inhabitants of dry lands [33].

4.3 Restoration initiatives

4.3.1 Baseline studies/surveys such as biodiversity assessments

Baseline surveys are studies that are done at the beginning of projects to collect information on project status before any types of intervention are implemented. Information obtained from such surveys later inform decision-makers on what impacts the projects have on target communities. For instance, the first step in protecting and managing biodiversity in any ecosystem is to understand what species exist by documenting these species and their environments through biodiversity surveys. Biodiversity is the biological variety and variability of life on Earth, a measure of variation at the genetic, species, and ecosystem levels.

Biodiversity performs multiple roles in the daily lives of people in the drylands through the supply of ecosystem services, food security, tourism, wealth creation and aiding a range of cultural services. As such, the value of Kenya's biodiversity resources cuts across the economic and social and, ultimately, the political pillars of Vision 2030. Direct benefits from plant diversity include food, medicine, honey, forage, vegetables, and other raw materials that play a vital role in the lives of poor people in rural and remote places [34]. Indirect benefits that flow from plant biodiversity's environmental services include employment, income, nitrogen fixation, maintenance of water cycles, regulation of climate, photosynthetic fixation of carbon dioxide, soil protection, storage, and cycling of essential nutrients as well as absorption and breakdown of pollutants [35]. Kenya's forests, for instance, play a vital role in rural livelihoods by providing food and energy for domestic consumption and watershed regulation. In Kitui and Mwingi districts of Kenya, plant species recorded during a survey of five hilltop sites were noted as important sources of medicine, fiber, food, fodder or forage, timber, and fuelwood [36]. Some of the much-sought species, especially for medicine, included *Warburgia ugandensis*, *Pittosporum viridiflorum*, *Securidaca longipedunculata*, *Zanthoxylum*, and *Strychnos*.

Overall, Kenya's known floral biodiversity assets include 7000 plants among other life forms [37]. However, the status of plant species' diversity in the dry lands is poorly documented [3]. In order to remain ahead in efforts to conserve dryland ecosystems, restore degraded sites, and reverse or halt desertification, the Government of Kenya gathers information from various sources to understand the prevailing circumstances and prescribe restoration measures using various tools and strategies. For instance, a study conducted in the Mutomo district, Kenya, evaluated the nature of degradation caused by selective logging for charcoal production and provided information on how this could be addressed to ensure the woodlands recovery without impacting negatively on the producers' livelihoods [38]. Restoring the dryland landscapes can help mitigate climate change, support sustainable livelihoods, and maintain biodiversity. Restoration aims to reestablish a previous ecosystem state and all its functions and services, while rehabilitation seeks to repair specific parts of the systems, in order to regain ecosystem productivity [39]. Effective restoration and rehabilitation of degraded drylands require a combination of policies and technologies and the close involvement of local communities. There are two categories of landscape-restoration methods that are commonly applied in the drylands: active restoration and passive restoration [40, 41].

4.3.2 Active-versus passive-restoration approaches

4.3.2.1 Active-restoration methods

These are methods involving management techniques, such as planting of seeds or seedlings. They are needed to restore severely degraded lands and are particularly relevant for water-limited environments, where self-restoration processes of severely degraded lands may be limited. One such method is the Framework Species Approach developed in Queensland, Australia [42]. The method involves a single planting of both early and late successional species. Planted species must survive in the harsh conditions of an open site as well as fulfill the functions of (i) fast growth of a broad dense canopy to shade out weeds and reduce the chance of forest fire and (ii) early production of flowers or fleshy fruits to attract seed dispersers and kickstart animal-mediated seed distribution to the degraded site [43]. Framework trees are indigenous, non-domesticated, forest tree species, which, when planted on deforested land, help to reestablish the natural mechanisms of forest regeneration and accelerate biodiversity recovery. A rainforest-restoration experiment established on abandoned pasture in northeastern Queensland to examine the effectiveness of five different restoration-planting frameworks concluded that some restoration success measures increased with planting diversity, but overall the rate of recovery was similar in framework species and maximum diversity method [44].

Active restorations, using afforestation and reforestation methods, are effective biological approaches with the potential to help restore and rehabilitate degraded dryland ecosystems and halt desertification [45]. Among other benefits, rehabilitation improves the soil biological activities where high rates of soil organic matter, organic C and N, suitable soil acidity range, and abundance of forest litter are considered the predisposing factors promoting higher microbial populations in enrichment planting as compared to secondary forest [46].

Afforestation is the planting of forests on land that historically had no forests [47]. The main purpose of afforestation is to reduce soil erosion by planting trees, which increases soil stability and enables forest regrowth. Other purposes include improving the potential wood extraction in the future and improving the visual landscape. However, there have been concerns that conversion of "natural" drylands to dryland forestry may have adverse ecological and environmental impacts on the environment, thus risking a wide range of ecosystem functions and services. Attempts have been made to demonstrate the potentially adverse implications of dryland forestry and highlight the caution needed when planning and establishing such systems [45]. For instance, in order to negate suppression of understory vegetation and sustain plant species' richness and diversity, low-density savanization by non-allelopathic tree species is preferred over high-density forestry systems by allelopathic species. According to the author, and wherever possible, it is preferable to plant native tree species rather than introduced or exotic species, in order to prevent genetic pollution and species invasion. In addition, mixed-species forestry systems should be favored over single-species plantations, as they are less susceptible to infestation by pests and diseases. In addition, drought-tolerant, fire-resistant, and less-flammable tree species should be preferred over drought-prone, fire-susceptible, and more flammable species.

4.3.2.2 Passive-restoration methods

These are methods in which no action is taken except to cease environmental stressors, such as agriculture or grazing, and are effective for restoring moderately degraded lands. In Eastern Kenya, for instance, the results of a study involving passive restoration show that woodlands have a high potential to recover if put under a suitable management regime since they have a high number of saplings [12]. The most commonly used of these approaches is the Assisted Natural Regeneration (ANR) approach, which acts on natural regenerates that are already present in deforested sites. The word "assist" in ANR refers to helping the naturally growing young trees to grow faster [48]. Assisted Natural Regeneration accelerates the natural succession process by protecting against disturbances, such as fire, stray domestic animals, and humans, and by reducing competition from grasses, bushes, and vines that would hinder the growth of naturally regenerated trees [48]. Forest restoration using ANR has advantages over conventional reforestation through planting by being cheaper to implement as costs associated with seedling production, site preparation, and planting are greatly reduced. The plant community that is established is well adapted to the site conditions, and the naturally regenerating plant community typically comprises a mixture of species that result in more diverse, multilayered vegetative cover [48].

Assisted Natural Regeneration is a flexible and adaptable approach that can be applied in a variety of contexts. It can, for instance, be combined with enrichment planting for various reasons including to fill in patches that may not have enough wildlings to establish tree canopy cover within the desired time frame, enhance the success and quality of forest restoration, and restore ecologically and/or economically valuable species to meet specific restoration objectives [48]. Enrichment planting may be defined as the introduction of valuable species to degraded forests without the elimination of valuable individuals who already existed at that particular site [49]. In this technique, trees are planted in gaps, lines, or open sites as plantations of mixed species or under canopies of young dryland forests. In a study to identify the optimal enrichment planting method vis-à-vis gap and line planting, and to evaluate the performance of two dipterocarps and three legumes planted in logged-over mixed deciduous forest of Laos, the diameter and height growth were favored more in gaps than in planting lines [50]. Furthermore, the use of logged-over gaps for enrichment planting was recommended given the difficulty to maintain constant line width and even light condition, the cost of annual clean operation, and the rigid geometric patterns of planting lines [50]. In Indonesia, gap planting with *Anthocephalus macrophyllus* to rehabilitate degraded natural forests increased soil density, although its value was categorized as a very loose soil class [51]. In another study from Malaysia, the total mean microbial enzymatic activity, as well as biomass carbon (C) and nitrogen (N) content, was significantly higher under enrichment planting than under secondary forest [46].

There is little consensus on whether active or passive restoration strategies are more successful for recovering biodiversity because few studies make adequate comparisons [41]. In some studies, recovery of species' richness and composition is similar in active- and passive-restoration sites, while in others, recovery of forest specialists is enhanced through active restoration [40, 41]. While both restoration strategies may lead to different vegetation structures, they may support similar biomass of foliage-dwelling arthropods and be similarly used by foraging insectivorous birds [40]. Passive restoration is generally less costly than active restoration and, if local and landscape characteristics do not impede recovery, may be a viable alternative. Where active restoration is adopted, it should be implemented using mixed plantations of native tree species and, whenever possible, select sites close to mature forest to accelerate the recovery of tropical forest biodiversity [41]. Because active restoration is more expensive than passive restoration, both strategies should be used in complementarity at the landscape level for cost-effectiveness and optimization of the different land management objectives for the wider landscape [40, 41].

4.3.3 Challenges of tree growing in drylands

Tree planting in the drylands poses challenges to land users, which are brought about by a combination of edaphic, ecological, and socioeconomic factors in these areas. These include, among others, moisture stress, termite infestation, animal damage, and competition from weeds [8, 52]. Although the farmers and tree growers have developed interest in tree planting as an investment activity, they are discouraged by the continuous low tree survival rate and thus are not able to reap the maximum benefits from their tree crop. In a guideline intended for farmers and tree growers living in the drylands of Kenya, the common factors contributing to tree mortality at all stages of tree growing have been presented [8]. They also provide interventions that can be applied during species' selection, raising seedlings in the nursery, out-planting, and tree management to enhance tree growing in the drylands. The Kenya Forestry Research Institute (KEFRI), through the dryland forest research program, has identified major factors contributing to low tree survival and developed mitigation measures, which include the selection of appropriate tree species and development of suitable methods for propagating, establishing, and managing trees [8]. KEFRI has demonstrated better ways of re-afforestation especially in areas under limited water availability [52], proposed species-site matching as a key consideration, especially with drought tolerance, and found mycorrhizal inoculation to greatly enhance the survival of trees planted in degraded areas, which have low mycorrhizal inoculum potential [53]. Inoculated trees have been used to restore the soil inoculum and to enhance the growth of interplanted agricultural crops. There are also opportunities to exploit tree/crop symbiotic associations in agroforestry systems, using trees selected both for their own attributes and for soil-improving qualities [53]. Farmers and tree growers need to adhere to these measures to improve tree survival and thus realize maximum profits from tree-planting activities.

4.3.4 Biodiversity policy and management in Kenya

4.3.4.1 Biodiversity law and policy

A number of legal and policy instruments have been put in place to enhance conservation and regulate utilization of biodiversity resources. Among these instruments are the Constitution of Kenya 2010, which entrenches a range of environmental imperatives and provides an avenue for remedying the land tenure, land use, and gender-inequity issues that have negatively affected the country's biodiversity. The Constitution also decentralized the management of a range of natural resources to the devolved units known as County Governments. Other biodiversity-related instruments include the Revised Kenya National Biodiversity Strategy and Action Plan (2010), Integrated Coastal Zone Management Policy (2010), Environment Management and Coordination Act (1999), National Water Policy (1999), Water Act (2002), Draft Forest Policy (2004), Draft ASALs Policy (2004), Forest Act (2005), Fisheries Policy (2008), Heritage Sites (2006), National Land Policy (2009), Energy Act (2006), Biodiversity Regulation (2006), Draft Wildlife Policy (2007), and the draft of Minerals and Mining Policy. Any initiative that directly or indirectly helps to conserve the country's biodiversity helps to meet the specific Vision 2030 poverty-alleviation objectives as well as the goal of improving the general welfare of citizens. A national biodiversity policy and law would be a useful complement to the above operative instruments.

4.3.4.2 Biodiversity research and development agenda for drylands

The Natural Resource Management (NRM) has been defined to mean *inter alia* the sustainable utilization of major natural resources, including forests, wild flora, and fauna [54]. Natural resources play an important role in providing fundamental life support, by proving a diversity of products and services, both social and ecological. Sustainable management of these resources is challenged by increasing demands, climate change, pollution, and economic development needs. These pressures have led to dwindling availability of natural resources, especially in the ASALs. The national research priorities that have been identified in Kenya to address the above challenges include, among others, the following [55]:

- Balance between productivity and environmental services
- Environmental protection for sustainable agriculture, livestock management, and aquaculture
- Studies on ecosystem services (including provisioning)
- Biodiversity and conservation of genetic resources

Biodiversity is also a key component of KEFRI's research and development agenda as articulated in the Institute's 2018–2022 Strategic Plan [56]. The current KEFRI Strategic Plan aims to achieve the following seven strategic objectives:

- Generating technologies for the establishment and management of forest plantations and trees on farms and enhance the production of superior germplasm for priority tree species for different agroecological zones
- Generating rehabilitation technologies for adaptation to climate change, sustainable forest landscapes, woodlands, wetlands, and riparian ecosystems
- Developing technologies for efficient processing and utilization of wood and non-wood forest products
- Formulating forestry policies for sustainable forest management and improved livelihoods
- Disseminating forestry research technologies and enhancing institutional research and development capacity
- Strengthening institutional capacity for research and development
- Enhancing corporate communication and publicity

KEFRI's specific actions in the development of technologies for rehabilitation and restoration of forests and allied natural resources in drylands target the following [56]:

- To develop guidelines on rehabilitation and restoration technologies and train stakeholders
- To establish permanent sample plots in forests and woodlands ecosystems for collection of data on ecological trends and dynamics
- To develop strategies for in-situ and ex-situ conservation for threatened and endangered species
- To perform ecological studies for various forest types to secure a broad range of goods and environmental services
- To develop technologies for sustainable natural forest and woodland management
- To quantify the impact of animal damage on forest ecosystems

4.3.5 Biodiversity conservation in Kenya

Although Kenya's biodiversity remains highly protected, declines are common phenomena due to a number of anthropogenic threats that have led to numerous conservation challenges [57]. Nevertheless, the country explores all avenues to ensure that efforts to win the war against biodiversity losses are sustained. Examples of progress made in research and development for drylands afforestation in Kenya include the following:

4.3.5.1 Selection of appropriate tree species

Until the 1980s–90s, when it became a government policy in Kenya to promote the replanting of indigenous rather than exotic tree species, most of the work on the selection of trees of arid and semiarid lands in Kenya was with exotic fast-growing species [53]. Since then, selection criteria have continued to evolve with consideration around preferences of local communities, availability of quality genetic material for propagation, and site biophysical conditions. Currently, a large number of tree species have been recommended for the drylands of Kenya [52, 53]. In the dryland areas of Kitui and Kibwezi (Eastern Kenya), tree species grown and recommended include *Azadirachta indica*, *Jatropha curcas*, *Senna siamea*, *Leucaena leucocephala*, *Croton megalocarpus*, *Casuarina equisetifolia*, *Melia volkensii*, *Eucalyptus camaldulensis*, and *Dovyalis caffra* [4].

In Kenya, the area under *Eucalyptus* is likely to increase as a result of high demand for transmission poles, for construction, fuelwood, carbon sequestration, and mitigation of the effects of climate change [58]. However, there is much unease about Eucalyptus water consumption as compared to other woody flora. On a positive note, studies have established that Eucalypts exhibit high efficiency in water use for biomass accumulation. It has been established that eucalyptus requires less water to produce one (1) kg of biomass than most crops [59].

Melia volkensii, an important timber species that grows well in well-drained soils, is a promising indigenous tree species found in the drylands of Kenya. It is fast growing, drought tolerant, and produces high-quality hardwood timber for furniture. However, this species is heavily exploited in its natural stands and the trend has been worsening over the last decade owing to shortage of alternative hardwood species in drylands. As a result, programs promoting domestication of the species as a plantation species are ongoing [60–62].

4.3.5.2 Development of suitable methods for propagation

Poor propagation of some promising tree species in the drylands of Kenya has slowed down the country's efforts to increase its forest cover to the targeted 10%. For example, lack of seedlings attributed to poor seed germination is experienced with *Terminalia brownii*, a drought-tolerant species, which can be used to rehabilitate degraded drylands through reforestation and agroforestry approaches [63]. The current demand for Terminalia seedlings is higher than the supply. Research has focused on development of technologies ranging from breakage of dormancy to plant-tissue cultures to improve germination propagation of such species. Studies have been conducted to investigate the dormancy and germination of *T. brownii* seeds collected from various dryland sites in Kitui, Makueni, Tharaka-Nithi, and Baringo Counties of Kenya. Extracted seeds recorded the highest germination with the best at 76% compared to nipped seeds (13%) and those subjected to other treatments [63]. *Melia volkensii* Gurke is another drought-tolerant tree native to the drylands, of which cultivation is limited by difficulties in propagation via conventional means. Full exploitation of the ability of thidiazuron to elicit regeneration in plant-tissue cultures, as a sole plant growth regulator, was found to be hampered by high costs. Alternative effective and low-cost agrochemical thidiazuron for in vitro propagation of *M. volkensii* was found to be Kingtai-TDZ, which has a high potency and suitability for use in tissue culture of the species [61]. Because of difficulties in seed germination, land users sometimes go for the use of plants produced from root and stem cuttings, rather than from seedlings, and researchers have focused on the possibility that root and stem cuttings may be used for propagation, rather than seedlings [62]. However, if cuttings are used to circumvent the problems of seed germination, alternative methods of controlling competition, such as root pruning, need to be considered.

4.3.5.3 Establishment and management of trees

In systems where trees are promoted for on-farm planting or in agroforestry systems, challenges associated with tree–crop interactions, and likely competition such as that for nutrient and water resources, are sometimes experienced. Researchers, for instance, find below-ground competition to be a major problem in simultaneous agroforestry systems and a focus of much research in recent decades [64]. Considering that trees raised from seed may differ in their competitiveness from those raised from cuttings, studies have been conducted to evaluate differences in root system architecture of plants raised from seed, stem, or root cuttings and the relationships between the competitivity index (CI) and crop yield [62]. From such studies, more shallowly rooted cuttings than seedlings, higher competitivity indices, and a negative relationship between CI and crop yield in *Melia volkensii* under integrated land-use systems have been observed. Therefore, to reduce tree–crop competition, the use of seedlings rather than cuttings should be recommended when promoting the use of this species on dryland farms [62].

4.3.5.4 Management of invasive plant species

Invasive plants are capable of penetrating and replacing the existing indigenous vegetation of a location [65]. These are mostly exotic plants that have been introduced in a location, either intentionally or unintentionally, and that reproduce and spread on their own [66]. In the late 1970s and the early 1980s, the East African dry lands witnessed the introduction of various alien species. They include the 10 key invasive plant species that affect the drylands of Kenya and Tanzania [65], namely *Lantana camara* (Lantana), *Prosopis juliflora* (Mesquite), *Prosopis pallida* (Mesquite), *Opuntia ficus-indica* (Prickly pear cactus), *Caesalpinia decapetala* (Mauritius thorn), *Psidium guajava* (Guava), *Senna spectabilis* (Cassia), *Acacia farnesiana* (Sweet acacia), *Acacia mearnsii* (Black wattle), and *Acacia polyacantha* (White thorn).

Invasive plants are a hazard in the tropical dry forests and rangelands of East Africa, having increasingly created disasters that have affected the environment and socioeconomic well-being of communities inhabiting these dry regions. Some of the negative effects of invasive species include causing the death of livestock by poisoning and destroying livestock foliage, accelerating biodiversity loss via suppression of native plants, and increasing diseases by offering a breeding ground for mosquitoes and other insects that carry ailments such as nagana and sleeping sickness [65]. Of the 10 invasive species identified in East Africa, 90% suppressed native plant species and reduced biodiversity. Because invasive species could cause food insecurity and slow economic growth, their potential to derail attainment of the Country's Vision 2030 targets cannot be underestimated [67]. In addition, there is a need to explore and exploit the range of livelihood opportunities that invasive alien species, such as *P. juliflora*, present in drylands. In this regard, KEFRI and other partners have, for many years, conducted research and developed technologies for the control and management of this species through utilization.

4.3.5.5 Extension services and outreach

Forestry extension is the art and science of converting information from research and past experiences to a practical level for use by local people who may not be specifically trained in forestry techniques [68]. There are two approaches to forestry extension, namely top-down approach and bottom-up/participatory approach. The bottom-up approach is a two-way information flow system that considers prior consultation with target beneficiaries about their needs/problems and aspirations for effective planning. The top-down approach is a one-way information delivery system that reinforces the hierarchical relationship between the extension agent and the client [69]. The aim of forestry extension in the drylands is to help the pastoral people manage their livelihoods and their environment and to involve them in forestry development activities. Although forestry activities remain the main concern of the forestry sector, pastoralists look at their management activities in totality. For this reason, an integrated approach to extension involving relevant National and County Ministries usually strengthens extension services in the ASALs.

The drylands of Kenya are extensive, with low productivity and sparse populations, thus calling for appropriate forestry extension methods. The extension techniques appropriate for the ASALs and methods of developing them have been outlined [68, 69]. Some of the most commonly used extension methods include exploratory seminars, small meetings, public meetings, individual visits, and field exercises [70]. Efforts are made to establish the two-way flow of information from the forestry extension agents to the target groups in order to explore local problems and their solutions in the context of local traditions, knowledge, needs, and priorities. Common extension tools used include the media (radio, film, television, print manuals, and posters), field demonstrations complemented by video and audio tapes, slide shows, local actors, and direct discussions.

5. Conclusion

This chapter provides an in-depth review of "desertification" as one of the most important challenges to livelihoods and development in the drylands of Kenya. The phenomenon results from climatic variations and human activities, such as destructions of water catchments and deforestation, that exert pressure on forests and woodlands leading to degradation, deforestation, and desertification. Desertification is, thus, a precursor to increased water scarcity. Overall, it causes reduced tree cover, reduces agricultural productivity, and increases water scarcity, climate change, destruction of water catchment areas and deforestation, which continue to increase water scarcity in Kenya. However, opportunities for remedial measures offer potential for intensified afforestation toward achieving the national target of 10% tree cover. More investment is needed for active and well-coordinated research toward rehabilitation and restoration of dryland resource systems. A significant impetus in combating desertification and drought is to devolve power to the people who are affected and to link environmental degradation to economic policies. Devolved action will attract local support to initiated programs. For sustainable development strategies to work, policies should put the welfare of the people in drylands at the center of the development agenda, uphold local people's rights, and empower the same people to adopt adaptive strategies to ascertain sustainable livelihoods. This review demonstrated the role of research and development in availing afforestation technologies in drylands to increase biodiversity and avert advancement of desertification in Kenya.

Acknowledgements

The authors acknowledge the support from the respective institutions of affiliations that allowed use of the data and information in their reports and granted us time to put together this write-up.

Conflict of interest

The authors declare no conflict of interest.

Author details

Joseph Hitimana[1*], Edward K. Mengich[2], Teresiah N. Kuria[1] and Pauline Kimani[3]

1 University of Kabianga, Kericho, Kenya

2 Kenya Forestry Research Institute, Lodwar, Kenya

3 ICAR-Indian Agricultural Research Institute, New Delhi, India

*Address all correspondence to: josephhitimana@kabianga.ac.ke

References

[1] Burrow E, Mogaka K. In: Behnke R, editor. Kenya's Drylands—Wastelands or an Undervalued National Economic Resource. Gland, Switzerland: IUCN; 2007. p. 44

[2] Mortimore M. Dryland Opportunities: A New Paradigm for People, Ecosystems and Development. IUCN, IIED, Gland, Switzerland: UNDP; 2009. p. 98

[3] Bonkoungou EG. Biodiversity in drylands: Challenges and opportunities for conservation and sustainable use. In: Maryam-Naimir-Fuller (UNDP/GEF), editor. The Global Drylands Partnership CIDA UNSO UNDP/GEF IIED IUCN WWF. NEF; Undated. Available from: https://www.iucn.org/sites/dev/files/content/documents/biodiversity-in-the-drylands-challenge-paper_0.pdf

[4] Kenya Forest Service. Forestry potential in the drylands. In: Forester: A Quarterly magazine of the Kenya Forest Service. Issue No. 12. April–June 2014

[5] Republic of Kenya. Economic Survey. Central Bureau of Statistics. Nairobi: Ministry of Planning and National Development; 2002

[6] Republic of Kenya. Sessional Paper on Sustainable Development of Arid and Semi-Arid Lands of Kenya (Draft). Nairobi: Office of the President; 2005. p. 30

[7] Darkoh MBK. The nature, causes and consequences of desertification in the drylands of Africa. Land Degradation and Development. 1998;**9**:1-20

[8] Mwamburi A, Musyoki J. Improving Tree Survival in the Drylands of Kenya: A Guide for Farmers and Tree Growers in the Drylands. KEFRI Information Bulletin No. 2. KEFRI, NALEP; 2010

[9] Carvajal-Escobar YM, Quintero-Angel M, Garcya-Vargas M. Women's role in adapting to climate change and variability. Advances in Geosciences. 2008;**14**:277-280

[10] Rafferty JP, Pimm SL. Desertification. In: Encyclopædia Britannica. 2019 [Accessed: 6 November 2019]

[11] Eswaran H, Lal R, Reich PF. Land degradation: An overview. In: Bridges EM, Hannam ID, Oldeman LR, Pening de Vries FWT, Scherr SJ, Sompatpanit S, editors. Responses to Land Degradation. Proc. 2nd. International Conference on Land Degradation and Desertification, Khon Kaen, Thailand. New Delhi, India: Oxford Press; 2001

[12] Ndegwa GM. Evaluating dry woodlands degradation and on-farm tree management in Kenyan drylands [PhD thesis]. Germany: University of Passau; 2017

[13] United Nations Convention to Combat Desertification. High-level dialogue on desertification, land degradation and drought. 2021. Available from: https://www.unccd.int/

[14] Trabucco A, Zomer R. Global Aridity Index and Potential Evapotranspiration (ET0) Climate Database v2. Figshare, Fileset. 2019. Available from: DOI:10.6084/m9.figshare.7504448.v3

[15] Trabucco A, Zomer RJ. Global Aridity Index and Potential Evapo-Transpiration (ET0) Climate Database v2. CGIAR Consortium for Spatial Information (CGIAR-CSI) [online]. 2018. Available from: https://cgiarcsi.community

[16] Wiesmeier M. Chapter 14: Environmental indicators of dryland. In: Environmental Indicators of Dryland Degradation and Desertification. The

Netherlands: Springer, Dordrecht; 2015. pp. 239-250

[17] https://www.conserve-energy-future.com/causes-effects-solutions-of-desertification.php

[18] WRI (World Resources Institute). Kenya GIS data [internet]. 2007. Available from: https://www.wri.org/data/kenya-gis-data [Accessed: 24 April 2021]

[19] Maina J, Wandiga S, Gyampoh B, Charles K. Assessment of land use and land cover change using GIS and remote sensing: A case study of Kieni, Central Kenya. Journal of Remote Sensing & GIS. 2020;**9**(1):1-5. DOI: 10.35248/2469-4134.20.9.270

[20] Admio AO, Njoroge JB, Claessens L, Wamocho LS. Land use and climate change adaptation strategies in Kenya. Mitigation and Adaptation Strategies for Global Change. 2012;**17**(2):153-171

[21] ESRI (Environmental Systems Research Institute). Kenya GIS data [internet]. 2011. Available from: https://www.arcgis.com/home/group.html?id=c3a358bb2438470f971fbfca9b2ff6d3&view=list#content [Accessed: 24 April 2021]

[22] Kipkorir J, Lang'at S. Variability of mangrove forests along the Kenyan coast. MARG I Final Report: WIOMSA-MARG I Contract No. 20/2007. KEMFRI; 2008

[23] Kinyanjui JM, Shisanya CA, Nyabuti KO, Waqo PW, Ojwala MA. Assessing tree species dominance along an agro-ecological gradient in the Mau Forest Complex, Kenya. Open Journal of Ecology. 2014;**4**:662-670

[24] Ogweno EDO, Opanga P, Obara AO. Forest landscape and Kenya's vision 2030. In: Proceedings of the 3rd Annual Forestry Society of Kenya (FSK) Conference and Annual General Meeting held at the Sunset Hotel, Kisumu; 30th September–3rd October, 2008. 2009

[25] Ototo G, Vlosky RP. Overview of the forest sector in Kenya. Forest Products Journal. 2018;**68**(1):6-14

[26] Kigomo BN. State of forest genetic resources in Kenya. Sub-Regional Workshop FAO/IPGRI/ICRAF on the conservation, management, sustainable utilization and enhancement of forest genetic resources in Sahelian and North-Sudanian Africa (Ouagadougou, Burkina Faso, 22-24 September 1998). In: Forest Genetic Resources Working Papers, Working Paper FGR/18E. Rome, Italy: Forestry Department, FAO; 2001

[27] Njoka J, Yanda P, Maganga F, Liwenga E, Kateka A, Henku A, Bavo C. Kenya: Country Situation Assessment Working Paper. 2016

[28] Chapter 7: Fresh water, coastal and marine resources. pp. 124-149. Available from: https://www.nema.go.ke/images/Docs/Regulations/KenyaSoECh7.pdf

[29] Nyunja J, Ochola S, Pengra B, Ochieng E. Kenya Wetlands Atlas. The Government of Kenya. Kenya: Ministry of Environment and Natural Resources; 2012

[30] Njoroge LW, Wahab AHA, Tracey SA, Oting WKA. Water resource in Kenya: Impact of climate change/urbanization. International Journal of Scientific and Research Publication. 2018;**8**(4):30-35

[31] Marshall S. The water crisis in Kenya: Causes, effects and solutions. Global Majority E-Journal. 2011;**2**(1):31-45

[32] Herlocker D, editor. Rangeland Resources in Eastern Africa: Their Ecology and Development. German Technical Cooperation (GTZ): Nairobi; 1999

[33] Republic of Kenya (Ministry of Environment and Natural Resources). National Action Programme A Framework For Combating Desertification in Kenya in the Context of the United Nations Convention to Combat Desertification. Nairobi: National Environment Secretariat; 2002

[34] UNEP. Global Environment Outlook—4: Environment for Development. Nairobi: United Nations Environment Programme (UNEP); 2007

[35] Kimenju JW, Kahangi EM, Rutto L, Mutua GK. Biotic constraints to banana production and the remedial measures adopted by farmers in Maragua District, Kenya. Journal of Applied Biosciences. 2010. Available from: http://erepository.uonbi.ac.ke:8080/xmlui/handle/123456789/52625

[36] Malonza PK, Muasya AM, Lange C, Webala P, Mulwa RK, Wasonga DV, et al. Final Report of Biodiversity Assessment in Dryland Hilltops of Kitui and Mwingi Districts. Nairobi, Kenya: RPSUD, NMK; 2006

[37] NEMA. National Environment Research Agenda for 2008-2030. Nairobi, Kenya: National Environment Management Authority (NEMA) and Government of Kenya; 2009

[38] Ndegwa GM, Nehren U, Grüninger F, Iiyama M, Anhuf D. Charcoal production through selective logging leads to degradation of dry woodlands: A case study from Mutomo District, Kenya. Journal of Arid Land. 2016;**8**:618-631. DOI: 10.1007/s40333-016-0124-6

[39] Evans J, Turnbull JW. Plantation Forestry in the Tropics. 3rd ed. Oxford: Oxford University Press; 2004

[40] Morrison EB, Lindell CA. Active or passive forest restoration? Assessing restoration alternatives with avian foraging behavior. Restoration Ecology. 2010;**19**(201):170-177

[41] Díaz-García JM, López-Barrera F, Pineda E, Toledo-Aceves T, Andresen E. Comparing the success of active and passive restoration in a tropical cloud forest landscape: A multi-taxa fauna approach. PLoS One. 2020;**15**(11):e0242020. DOI: 10.1371/journal.pone.0242020

[42] Steve G, Nigel IJT. Repairing the Rainforest: Theory and Practice of Rainforest Re-establishment in North Queensland's Wet Tropics. Queensland: Wet Tropics Management Authority; 1995

[43] Betts H. The framework species approach to forest restoration using functional traits as predictors of species performance [doctor of philosophy thesis]. University of Liverpool; 2013

[44] Singarayer KF, Pohlman CL, Westbrooke ME. The effectiveness of different planting frameworks for recruitment of tropical rainforest species on ex-rainforest land. Restoration Ecology. 2015;**24**(3):364-372. DOI: 10.1111/rec.12317

[45] Stavi I. Seeking environmental sustainability in dryland forestry. Forests. 2019;**10**(737):1-6. doi:10.3390/f10090737

[46] Karam DS, Arifin A, Radziah O, Shamshuddin J, Majid NM, Hazandy AH, et al. Impact of long-term forest enrichment planting on the biological status of soil in a deforested Dipterocarp Forest in Perak, Malaysia. The Scientific World Journal. 2012(2):641346. Available fom: https://doi.org/10.1100/2012/641346

[47] Millennium Ecosystem Assessment. Ecosystems and Human Well-being: Synthesis. Washington, DC: Island Press; 2005

[48] FAO. Restoring Forest Landscapes Through Assisted Natural Regeneration

(ANR)—A Practical Manual. Bangkok: Food & Agriculture Org; 2019. p. 52

[49] Montagnini F, Eibl B, Grance L, Maiocco D, Nozzi D. Enrichment planting in overexploited subtropical forests of the Paranaense region of Misiones, Argentina. Forest Ecology and Management. 1997;**99**:237-246

[50] Sovu, Tigabu M, Savadogo P, Oden PC, Xayvongsa L. Enrichment planting in a logged-over tropical mixed deciduous forest of Laos. Journal of Forestry Research. 2010;**21**:273-280

[51] Elias E, Suwarna U. Impacts of gap planting on soil density and erosion. Jurnal Penelitian Kehutanan Wallacea. 2019;**8**(1):9-18

[52] Kenya Forestry Research Institute. Tree Planting and Management Techniques under Limited Water Availability: Guideline for Farmers and Extension Agents. Kenya: KEFRI, JIFPRO; 2014

[53] Milimo PB, Dick JMP, Munro RC. Domestication of trees in semi-arid East Africa: The current situation. In: Leakey RRB, Newton AC, editors. Tropical Trees: Potential for Domestication, Rebuilding Forest Resources. London: HMSO; 1994. pp. 210-219

[54] Darby S. Natural resource governance: New frontiers in transparency and accountability: Transparency & accountability initiative. London. 2011. Available from: http://www.transparency-initiative.org/wpcontent/uploads/2011/05/natural_resources_final1.pdf

[55] Republic of Kenya (Ministry of Education). National Research Priorities 2018-2022. 2019. Available from: www.nacosti.go.ke [Accessed: June 2019]

[56] Kenya Forestry Research Institute. Strategic Plan 2018-2022. Nairobi: KEFRI; 2018

[57] Hitimana J, Ole Kiyiapi JL, Kibugi PW, Kisioh H, Mayienda R, Warinwa F, Lenaiyasa P, Sumba D. In: Grillo O, editor. Challenges of Linking Socio-Economic Significance and Conservation Value of Forests in Drylands of Kenya: Case Study of Kirisia Forest-Samburu Pastoralists Coexistence, Biological Diversity and Sustainable Resources Use. InTech; 2011. Available from: http://www.intechopen.com/books/biological-diversity-andsustainable-resources-use/challenges-of-linking-socio-economic-significance-and-conservation-value-offorests-in-drylands-of-k

[58] Albaugh JM, Dye PJ, King JS. Eucalyptus and water use in South Africa. International Journal of Forestry Research. 2013;**2013**:852540. DOI: 10.1155/2013/852540

[59] Munishi PKT. The eucalyptus controversy in Tanzania. In: Paper Presented at TAF Annual General Meeting (AGM) 23rd–24th 2007 Dodoma Tanzania; 2007

[60] Muok B, Mwamburi A, Kyalo E, Auka S. Growing *Melia volkensii*. A Guide for Farmers and Tree Growers in the Drylands. KEFRI Information Bulletin No. 3, Nairobi, Kenya; 2010

[61] Mulanda ES, Adero MO, Amugune NO, Akunda E, Kinyamario JI. High-frequency regeneration of the drought-tolerant tree *Melia volkensii* Gurke using low-cost agrochemical thidiazuron. Hindawi Publishing Corporation. Biotechnology Research International. 2012;**2012**:5. Available from: https://doi.org/10.1155/2012/818472

[62] Mulatya JM, Wilson J, Ong CK, Deans JD, Sprent JI. Root architecture of provenances, seedlings and cuttings of *Melia volkensii*: Implications for crop yield in dryland agroforestry. Agroforestry Systems. 2002;**56**:65-72

[63] Okeyo MM, Obwoyere GO, Makanji DL, Njuguna JW, Atieno J. Promotion of *Terminalia brownii* in reforestation by development of appropriate dormancy breaking and germination methods in drylands; Kenya. Global Ecology and Conservation. 2020;**23**:e01148. DOI: 10.1016/j.gecco.2020.e01148

[64] Van Noordwijk M, van de Geijn SC. Root, shoot and soil parameters required for process-oriented models of crop growth limited by water or nutrients. Plant and Soil. 1996;**183**:1-25

[65] Obiri JF. Invasive plant species and their disaster-effects in dry tropical forests and rangelands of Kenya and Tanzania. Jamba: Journal of Disaster Risk Studies. 2011;**3**:417-428. DOI: 10.4102/jamba.v3i2.39

[66] Rejmanek M. What makes a species invasive? In: Pysek P, Prach K, Rejmanek M, Wade M, editors. Plant Invasions—General Aspects and Special Problems. Amsterdam: SPB Academic; 1995. pp. 3-13

[67] Howard GW, Matindi SW. Alien invasive species in Africa's wetlands—Some threats and solutions. In: IUCN-The World Conservation Union, The Ramsar Convention on Wetlands, and The Global Invasive Species Programme. Nairobi: IUCN—The World Conservation Union Regional Office for Eastern Africa; 2003

[68] Peace Corps. Reforestation in the Pacific Islands. Information Collection and Exchange M0033; 1990

[69] Ahmed MR. Extension for community development: Planning a forestry extension programme. In: Paper Submitted to the XII World Forestry Congress. Quebec City, Canada; 2009

[70] Kenya Forestry Research Institute. A Dryland Forestry Handbook for Kenya. Nairobi, Kenya: KEFRI; 1992. p. 95

Chapter 12

Desertification and Its Control along the Qinghai-Tibet Railway

Yuguo Liu, Jiufu Luo, Jinxing Zhou and Ming Cui

Abstract

The Qinghai-Tibet Railway is a magnificent project in the twenty first century. However, the problem of land desertification has arisen during the operation of the railway. Many sections of the railway roadbed are buried by sand. The ecological safety along the railway and the safe operation of the railway have attracted worldwide attention. This chapter will focus on the current situation of desertification along the Qinghai-Tibet Railway, such as key desertification sections and the temporal and spatial characteristics of the occurrence of desertification. At the same time, it introduces the characteristics of the dynamic conditions of railway desertification and the source of sand material. It is divided into two parts: biological measures and engineering measures to introduce desertification control along the railway. The biological measures focus on the selection of *Lolium perenne*, *Festuca sinensi*, *Elymus breviaristatus*, *Elymus nutans* and *Poa crymophila,* and other alpine native sand-fixing plant materials. The engineering measures will introduce the railway desertification comprehensive prevention and control technology system that combines solidification, resistance, and transportation.

Keywords: Qinghai-Tibet, railway, desertification, sand prevention and control, biological measures

1. Introduction

The Qinghai-Tibet Plateau is the highest and unique physical geographic unit in the world. It has a significant impact on the ecological environment of China and its neighboring countries. For a long time, poor traffic conditions in the plateau area have greatly restricted the overall economic and social development of the region. The Golmud-Lhasa section of the Qinghai-Tibet Railway started construction in July 2001 and opened to traffic on July 1, 2006. The Qinghai-Tibet Railway project has fundamentally changed the traffic conditions in the Qinghai-Tibet Plateau, and promoted the reform and opening up of the region, the overall economic and social development, and the improvement of the living standards of the Tibetan people.

The Qinghai-Tibet Plateau has the characteristics of high altitude, low temperature, little precipitation, simple ecosystem structure, weak anti-interference ability, and vulnerability to global environmental changes, showing strong vulnerability [1]. The permafrost section of the Qinghai-Tibet Railway is about 550 km, and the section with an altitude of more than 4000 m is 960 km. The terrain and landforms in the line area are changeable, with high altitude, deep-frozen soil, strong winter wind, low temperature, and large temperature difference between day and night. Plants are difficult to root and have a short growth period [2]. The plants only have a growth time of about 3 months. In addition, in recent years, due to the increasing global climate change and human disturbance, the trend of vegetation degradation is obvious. Once the vegetation in this area is destroyed, it is very difficult to restore [3]. The degraded grassland has become the source of railway sand-damaged materials [1]. In the long-term investigation and monitoring, it was discovered that the construction of the railway has changed the wind and sand environment along the line so that the wind and sand that could have passed through the border piled up near the roadbed. Sand accumulation on the trackbed directly affects the safe operation of railways [4]. At the same time, the accumulation of sand near the roadbed interferes with the protection measures of the railway's frozen soil, accelerated the melting of the frozen soil under the roadbed, caused the deformation of the roadbed, and caused greater harm to the safe operation of the railway. The sand disaster has seriously threatened the operational safety of the Qinghai-Tibet Railway, and it is one of the most severe sand-damaged lines in China [5].

However, there are still huge challenges in understanding the formation mechanism and distribution pattern of sand damage in the Golmud-Lhasa section of the Qinghai-Tibet Railway. Sand hazard prevention and control technologies were also just in their infancy. In the early stage of railway construction, the breeding and selection of native herbaceous plants used for vegetation restoration along the railway have been carried out, and certain results have been achieved. However, woody plants suitable for cold and dry environments have been extremely scarce. The supporting technologies for suitable artificial vegetation construction were insufficient. The traditional biological measures for sand prevention and control along the railway and the "tree, shrub, and grass" model are difficult to achieve. The existing sand prevention technologies along the railway have a series of problems such as high engineering cost, low efficiency, lack of materials required for biological measures, and imperfect comprehensive sand prevention and control system.

On the whole, in order for the Qinghai-Tibet Railway to become an environmentally friendly and safe railway, it is necessary to understand the distribution and characteristics of the severe and potential sand-damaged areas along the railway and strengthen the research on the mechanism and comprehensive prevention and control technology of sand damage along the railway. Urgent problems to be solved include degradation mechanism and protection technology of native vegetation around railways, rapid vegetation restoration technology, railway protection forest plant selection and planting supporting technology, and the combination of biology and engineering technology, etc. This chapter will introduce the distribution pattern and cause mechanism of sand damage, the research and development and integration of vegetation restoration technology, and the research and integration of engineer and plant sand control technology along the Qinghai-Tibet Railway.

2. The distribution pattern and cause mechanism of sand disasters along the Qinghai-Tibet Railway

In terms of space, the sand disasters along the Qinghai-Tibet Railway are mainly distributed from Golmud (K815 + 380) to Tibet's Cuona Lake (K1531 + 280). The sand damage in this section is not evenly distributed, mainly concentrated in the Hongliang River (K1104 + 690), Tuotuo River (K1224 + 810), Za'gya Zangbo (K1445 + 560) river valleys, and the areas on both sides of the river and Cuona Lake section (K1528 + 710 to K1531 + 280) and other areas [6]. The railway section with severe sand damage is 78.8 km (**Figure 1**).

In terms of time, the Qinghai-Tibet Railway has many windy days in winter and spring. Especially the main plateau, including Tuotuo River, Wudaoliang, Amdo,

Figure 1.
Sand disasters along the Qinghai-Tibet Railway. (a) Hongliang River section, (b) Tuotuo River section, (c) Za'gya Zangbo section, (d) Cuona Lake section.

and other areas, the number of strong wind days is more than 100 days. Moreover, the climate is relatively dry in winter and spring. For example, in the Tuotuo River area, according to the observation data of the weather station for many years, the average annual wind speed is greater than 4 m s^{-1}. The number of windy days is more than 140 days. The number of sandstorm days in the year is 15–22 days. The annual precipitation is about 200 mm, mainly from June to September (about 85%). There is no snow cover in winter and spring, and the surface is extremely dry. At the same time, this period is also a period of high winds, with the highest wind speed reaching 32 m s^{-1}. Therefore, the Qinghai-Tibet Railway sand disaster occurred in these seasons.

The main body of the Qinghai-Tibet Railway is located in the rapids area in the middle of the westerly belt. Observation of sandstorms on the Hongliang River, Tuotuo River, Cuona Lake, and other road sections with severe sand damage revealed that the wind along the railway is strong and the amount of sand transported is relatively large. The sanding wind has a single wind direction and a long duration. It is dominated by westerly winds. These conditions have provided sufficient impetus for wind-sand activities [6]. The sand materials along the Qinghai-Tibet Railway mainly come from river sediments, desertified meadows and grasslands, and lake sediments. River facies sediments are mainly concentrated in river valley areas. Affected by topography and wind, sand materials develop from the river valley area to the two banks. The wind-sand disasters are particularly serious on the downwind bank of the river valley. This type of sand material is mainly distributed in the valleys and banks of the Hongliang River, the Xiushui River, the Beilu River, the Tongtian River, the Tuotuo River, the Za'gya Zangbo, and the Basuoqu River. Sandy meadows and grassland sand sources are typical non-point source sand sources, which are widely distributed. The main damage area is the Tuotuo River section. The impact of sandy meadows and grassland sand sources on railways mainly has two aspects. On the one hand, under favorable wind conditions, sand particles accumulate on the roads and directly harm the railway. On the other hand, sand particles enter the river valley with water flow, and deposit in the river valley area, becoming an important source of river facies sediments. Lake sediments are also an important source of sand material along the Qinghai-Tibet Railway, especially the Cuona Lake section, which has also become the most severe sandstorm area along the Qinghai-Tibet Railway. In addition, the sources of sand along the Qinghai-Tibet Railway also include rock weathered debris, sandy Gobi, wind erosion of ancient dunes, and activation of fixed dunes. The sources of these wind-sand disasters either directly harm the railways, or compound with each other, and superimposed on the railways.

After the completion of the Qinghai-Tibet Railway, due to the appearance of the roadbed, the original relatively stable dynamic balance of the plateau sand movement was disturbed spatially, and the flow field structure and transportation and accumulation conditions of the near-surface sand flow were changed, resulting in the deposition of sand materials near the railway. The hazards of sandstorms were highlighted. Through the wind tunnel simulation experiment on the characteristics of the flow field of the Tuotuo River section of the Qinghai-Tibet Railway, the formation mechanism of the sand damage of the roadbed was studied in combination with the flow field structure on both sides of the roadbed and the horizontal gradient distribution of the wind speed profile. It was found that when the airflow passed through the railway subgrade, there were obvious obstructed uplift areas, current gathering acceleration areas, deceleration and subsidence areas, and dissipation and recovery areas. The railway subgrade affected the characteristics of wind-sand flow by changing the movement state of the airflow, the separation of the boundary layer, and the size of the return zone. The formation mechanism of railway sand hazards was mainly determined by the functional zones where the air currents were located on both sides of the railway. When the wind-sand flow run near the railway subgrade, as the airflow encountered obstacles and rises, energy consumption was large, the wind speed of the bottom airflow decreased, and sand particles accumulated at the foot of the windward slope of the railway in the way of falling and depositing, causing sand burial on the railway subgrade. The airflow on the windward side of the subgrade mainly caused wind erosion to the middle of the subgrade or the shoulder of the subgrade due to the uplift and the acceleration of the current collection due to obstacles. When the airflow crossed the railroad track on the leeward side due to decelerating settlement and vortex movement, the sand carrying capacity was drastically reduced, and the sand flow was in a state of supersaturation, which will inevitably accumulate a large amount of sand particles carried on the leeward slope [7, 8].

3. Vegetation restoration along the Qinghai-Tibet Railway

The vast majority of areas where the Qinghai-Tibet Railway crosses are of grassland vegetation, mainly distributed from Xidatan to Yangbajing. The process of railway construction and operation greatly affects the grassland ecosystem along the route. The grassland ecosystem along the railway is aging and degraded. In addition, during the construction of the railway, engineering sites, such as borrow pits and roadbed slopes have a huge impact on the grassland ecosystem along the line. The research and development of vegetation restoration technologies for grasslands of different site types is imminent. Based on this, we screened suitable plant species for difficult sites such as degraded alpine grasslands, roadbed slopes, and borrow sites to establish a stable and healthy plant community [9, 10].

3.1 Vegetation restoration on the railway construction affected land

Before vegetation restoration is carried out on the borrow site, the area to be restored shall be prepared. In areas with greater impact on the landscape, high requirements for re-vegetation, and topsoil coverage, topsoil backfilling can be implemented, and the soil can be modified by applying biological fertilizers. In areas with a general impact on the landscape, mature sheep dung and yak dung can be used to improve the soil. At the same time, when sowing the seeds of perennial herbaceous plants, mix with seed base fertilizer. For the bare land in the arid section of the alpine grassland in the north of the Tanggula Mountain of the Qinghai-Tibet Railway, *Elymus nutans*, *Poa crymophila*, *Leymus scalenus*, and *Roegneria thoroldiana* can be selected as the main plants to regenerate alpine vegetation. For the bare land of the low-lying saline-alkali area, *Puccinellia distans*, *Poa crymophila*, *E. nutans*, *Leymus chinensis*, and *Puccinellia tenuiflora* are the main plants for rebuilding alpine vegetation. For the bare land between Tanggula Mountain Pass and Amdo in the alpine meadow area south of Tanggula Mountain, vegetation restoration should focus on topography and soil improvement. Use the natural succession of vegetation to restore the natural vegetation of the soil taking and spoiling ground. But for borrow sites that are close to the railway line and affect the landscape, *E. nutans*, *Elymus sylvestris*, *E. dahuricus*, etc. can be planted. For bare lands such as the soil removal field between Amdo and Sangxiongla Mountain Pass along the railway, *E. nutans* can be selected as the main regenerated grass species. For the bare land from the Sangxiongla mountain pass to Yangbajing along the railway, *E. nutans* and *E. sylvestris* can be selected as the main grass species for alpine meadow regeneration and can be matched with *Elymus dauri* and auxiliary grass species such as *P. tenuiflora*, *Leymus sativus*, *Festuca arundinacea*, and *Bromus inermis* [1, 11].

3.2 Vegetation restoration in desert areas

The desert area is mainly distributed in the section from Golmud to Kunlun Mountain. The average soil moisture content in this section is extremely low, less than 4%. Vegetation coverage is very low, except for a few areas where vegetation coverage is higher, often less than 30%. The species richness is also very low. Land desertification is serious. Super xerophytic shrubs and semi-shrubs grow sporadically in some areas. This is the result of an extremely difficult habitat that has long been adapted to dry climates and severe water shortages. The drought-tolerant species of Chenopodiaceae, Tamaricaceae, Zygophyllaceae, and other drought-tolerant plants form the arid and semi-arid desert vegetation landscape. For vegetation restoration in this area, shrubs such as *Sympegma regelii*, *Calligonum mongolicum*, *Salsola abrotanoides*, *Nitraria tangutorum*, *Ceratoides latens*, *Ephedra przewalskii,* and *Oxytropis aciphylla* can be selected. Among them, *S. regelii* is suitable for mild saline-alkaline deserts, *C. mongolicum* is suitable for gravel deserts, *S. abrotanoides* is suitable for gravel deserts, and *C. latens* is suitable for Gobi and desert areas.

3.3 Vegetation restoration in alpine grassland and alpine meadow

The alpine grassland area is mainly distributed between Xidatan and Fenghuoshan. The total coverage of natural vegetation is between 40% and 60%. Cold and drought-tolerant herbs such as *Stipa purpurea*, *Elymus nutans*, *Poa poophagorum*, *Carex moorcroftii*, *Saussurea arenaria*, *Leontopodium pusillum*, *Kobresia robusta*, *Ajania przewalskii*, *Littledalea racemosa*, *Roegneria thoroldiana*, etc. can be used. The alpine meadow area is mainly distributed from the north of Tanggula

Mountain to Jiuzina Peak. In the case of not being destroyed, the total coverage of natural vegetation is often 100%. Plants available for this segment include *Kobresia pygmaea*, *Kobresia littledalei*, *Kobresia humilis*, *Poa* spp., and *Festuca* spp. plants.

3.4 Excellent herbaceous plants tolerant to sand burial

Sand materials accumulate in the area along the Qinghai-Tibet Railway with severe sand damage. Ordinary plant species are easily buried by sand, causing vegetation restoration to fail. Sand burial resistance test was carried out on five native herbaceous plants: *Lolium perenne*, *Festuca sinensi*, *Elymus breviaristatus*, *Elymus nutans* and *Poa crymophila* [12].

The depth of sand burial had a significant effect on the plant height and root length of 5 grasses. At the same time, seeds with a high thousand-grain weight have a high germination rate. At a burial depth of 0–6 cm, the plant height of *L. perenne* showed an upward trend and then a downward trend, while the root length showed an increasing trend. The maximum plant height and maximum root length appeared at the buried depth of 4 cm and 6 cm. The plant height and root length of *F. sinensi*, *E. breviaristatus* and *E. nutans* all showed a trend of first increasing and then decreasing. For the maximum plant height and root length, *F. sinensi* appeared at 1 cm and 2 cm burial depths, *E. breviaristatus* appeared at 3 cm burial depths, and *E. nutans* appeared at 2 cm and 3 cm burial depths. However, *P. crymophila* failed to emerge at a depth of more than 3 cm in the sand. The maximum plant height appears at a burial depth of 0–1 cm, and the maximum root length appears at a burial depth of 2 cm.

Therefore, in the process of vegetation restoration and reconstruction, seeds with good quality and high thousand-grain weight should be selected for the restoration of moderate sand-buried vegetation with a thickness of 1–3 cm in order to increase the rate of seed emergence and increase the rate of seedling colonization. At the same time, *L. perenne*, *F. sinensi*, *E. breviaristatus* and *E. nutans* can be used for vegetation restoration in areas with severe sand damage along the line. *Poa crymophila* can be used for vegetation restoration in mild sand-damaged areas where the accumulation rate of sand material is low.

4. Sand prevention and control technology along the Qinghai-Tibet Railway

The frequent occurrence of sand disasters along the Qinghai-Tibet Railway has seriously threatened the safe operation of the railway. For a long time, the prevention and control of sand disasters along the Qinghai-Tibet Railway have mostly adopted the management model of "distant resistance and adjacent fixation". That is, high vertical sand barriers such as sand retaining walls are built in sections far away from the railway to block wind and sand flow, and low vertical sand barriers such as stone grids are built to fix sand surfaces in sections near the railway [13]. But in fact, this governance model does not scientifically determine the source of wind and sand and the path of wind and sand movement. The function of the sand retaining wall to block the wind and sand flow is not fully utilized. The stone grid sand barrier is often buried by the wind and sand and loses its effectiveness. A large amount of accumulated sand has posed a huge threat to railway tracks and roadbeds [14]. Therefore, the Qinghai-Tibet Railway Company needs to spend a lot of manpower, material resources, and financial resources for manual and mechanical sand removal in the road area every year. In addition, due to the lack of sand barrier construction materials such as stone grids in some areas, has objectively pushed up the cost of prevention and control of sandstorms on the Qinghai-Tibet Railway.

In view of this, on the basis of scientifically determining the characteristics and mechanism of sand damage along the Qinghai-Tibet Railway, and fully combining the natural conditions and the degree of land desertification along the Qinghai-Tibet Railway, the wind-sand disaster prevention system along the Qinghai-Tibet Railway has been explored and developed. It broke the traditional governance model of "distant resistance and adjacent fixation". The system mainly includes: (1) proposed the "source control, comprehensive prevention and control" governance technology model. By clarifying the source and path of sand materials, the protection and restoration of the source can be carried out to achieve source control. (2) Developed a governance technology system that combines "fixation, resistance, and transmission". Based on the site conditions of sand-damaged sections and the suitability of sand-fixing materials, a series of "sand-fixing" technologies suitable for use along the Qinghai-Tibet Railway has been developed, including the use of soil modification and mulching technology, planting bag technology, and multi-grass mixed planting technology for rapid vegetation reconstruction and rapid sand fixation technology using environmentally friendly chemical sand fixation agents. The "blocking" sand technology mainly refers to the use of new High-density polyethylene (HDPE) sand barrier materials. By comparing the single-width sand transport under the conditions of natural sand transport and HDPE sand fixation barriers, the sand control efficiency of HDPE below 1 m from the ground surface can reach 67.82%, of which the sand control efficiency can reach 86.34% below 20 cm from the ground surface. The sand transport technology is mainly to invent the sand transport railway subgrade. It is mainly suitable for the severe sandstorm section of railway embankment where the main wind direction is single and the angle is large or perpendicular to the railway line, in order to reduce the threat of sandstorm. (3) Invented the technology of combining biological sand fixation with engineering sand control. It combined high vertical sand barriers, cement board grid sand barriers and vegetation restoration measures. The cold-resistant, drought-resistant, and sand-tolerant native plants selected by the project were used as vegetation restoration materials, including Elymus, Poa, and Fescue spices [15, 16].

Figure 2.
Sand prevention and control technology along the Qinghai-Tibet Railway. (a) The promotion and application of the concept of "sand fixing in the distance". (b) Biological measures + engineering measures realize the control of sand materials from the source. (c) Low vertical net sand barriers are combined with biological measures to control sand sources. (d) High vertical net sand barriers are combined with biological measures to control sand sources.

The above-mentioned sand disaster control technology has been demonstrated and promoted in the severely sand-hazard sections of the Qinghai-Tibet Railway such as Hongliang River, Xiushui River, Beilu River, Tuotuo River, Tongtian River, Za'gya Zangbo, Cuona Lake, etc (**Figure 2**). The Qinghai-Tibet Railway Company adopted it in the prevention and control of the 78.8 km railway section of severe sand damage. It overcomes the shortcomings of using long-distance resistance and adjacent fixation in the past to intercept sand near the railway subgrade. This has effectively curbed the occurrence of railway sand disasters. Thus, the safe operation of the Qinghai-Tibet Railway is ensured.

5. Conclusion

With the operation of the Qinghai-Tibet Railway, severe wind and sand disasters gradually appeared along the line, threatening the safety of the railway. There is 78.8 km of severe sand-damaged roads along the route, distributed in 7 sections from Golmud to Nanshankou, Hongliang River, Xiushui River, Beilu River, Tuotuo River, Southern of Tanggula Mountain to Za'gya Zangbo, and Cuona Lake. Sand disasters mainly occurred from November to April of the following year. "Wind and drought in the same season" makes the wind and sand activities have sufficient power conditions and material basis. The airflow encounters resistance and rises on the windward side and slows down and settles on the leeward side, resulting in the accumulation of sand on both sides of the railway subgrade. The uplift of the airflow and the acceleration of the current collection cause wind erosion in the middle of the roadbed and the road shoulder. The low clearance height of the bridge caused sand accumulation on both sides of the railway bridge. The sand along the line is rich in materials and has complex sources. The main sources of sand include river and lake sediments, desertified meadows, and grasslands. Based on the distribution pattern and cause mechanism of sand damage along the railway, the Qinghai-Tibet Railway Sand Hazard Integrated Prevention and Control Technology System, which combines engineering and biology, and solidification, resistance, and transportation, can effectively prevent the occurrence of railway sand damage, thereby ensuring safe railway operation.

Acknowledgements

There are many scientific researchers studying sand disasters along the Qinghai-Tibet Railway. Our article is a brief introduction to the overall work. There are many people in the author who cannot be listed, including Jianjun Qu, Jingming Zheng, Yuzhou Ke, Shengbo Xie, Jinchang Wang, Kun Guo, and Guanglei Gao. Thank them for their contributions.

Author details

Yuguo Liu[1]*, Jiufu Luo[2], Jinxing Zhou[2] and Ming Cui[1]

1 Institute of Desertification Studies, Chinese Academy of Forestry, Beijing, China

2 Beijing Forestry University, Beijing, China

*Address all correspondence to: liuyuguo@caf.ac.cn

References

[1] Zhou J, Yang J, Peng G. Constructing a green railway on the Tibet plateau: evaluating the effectiveness of mitigation measures. Transportation Research Part D. 2008;**13**(6):369-376

[2] Yin G, Niu F, Lin Z, et al. The distribution characteristics of permafrost along the Qinghai-Tibet railway and their response to environmental change. Journal of Glaciology and Geocryology. 2014; **36**(4):772-781

[3] Wang Z, Zhang Y, Yang Y, et al. Quantitative assess the driving forces on the grassland degradation in the Qinghai–Tibet Plateau, in China. Ecological Informatics. 2016;**33**:32-44

[4] Xie S, Qu J, Pang Y. Dynamic wind differences in the formation of sand hazards at high- and low-altitude railway sections. Journal of Wind Engineering and Industrial Aerodynamics. 2017;**169**:39-46

[5] Xie S, Qu J, Wang T. Wind tunnel simulation of the effects of freeze-thaw cycles on soil erosion in the Qinghai-Tibet plateau. Sciences in Cold and Arid Regions. 2016;**8**(3):187-195

[6] Zhang K, Qu J, Yao Z, et al. Sand damage and its control along the Golha section of the Qinghai-Tibet railway. Arid Land Geography. 2014;**37**(1):74-80

[7] Yao Z, Qu J. Source and grain size of aeolian sands along Golmud-Lhasa section of Qinghai-Tibet railway. Journal of Desert Research. 2012;**32**(2):300-307

[8] Zhao Y, Gao G, Zhang Y, et al. Aeolian sediment fingerprinting in the Cuona Lake section along the Qinghai-Tibetan railway. Journal of Cleaner Production. 2020;**261**:121233

[9] Luo J, Zheng J, Zhou J, et al. Analysis of the interspecific associations present in an alpine meadow community undergoing revegetation on the railway-construction affected land of the Qinghai-Tibet plateau. Acta Ecologica Sinica. 2016;**36**(20):6528-6537

[10] Zhou L, Xu Z, Zhou G, et al. Mechanics strength test of herb roots on the earth-borrowed area of Tuotuo River section of the Qinghai-Tibet railway. Arid Zone Research. 2020; **37**(5):1353-1361

[11] Wei J. Study of vegetation restoration and revegetation during the constructing of the Qinghai-Tibet railway. Journal of Glaciology and Geocryology. 2003;**25**(S1):195-198

[12] Guo K, Ren K, Zheng J, et al. Effect of sand burial depth on seed germination and seedling growth of four grasses in Qinghai-Tibet plateau. Journal of Fujian Agriculture and Forestry University (Natural Science Edition). 2018;**47**(4):487-493

[13] Xie S, Qu J, Pang Y, et al. Causes and controlling mode of sand hazards in Honglianghe section of Qinghai-Tibet railway. Journal of the China Railway Society. 2014;**36**(11):99-105

[14] Jiang X, Zhou J, Liu Y, et al. Effectiveness of sand control measures in Nanshankou area along Qinghai-Tibet railway. Scientia Silvae Sinicae. 2016; **52**(10):55-63

[15] Niu Q, Qu J, Zhang K, et al. Status of aeolian-sand disaster and estimation of mechanical sand-controlling benefit at typical sections of Qinghai-Tibet railway. Journal of Desert Research. 2009;**29**(4):596-603

[16] Luo J, Liu X, Deng D, et al. Film mulched co-applied with compost improves revegetation in sand dunes fixed by cement checkerboard along the Qinghai-Tibet railway. Fresenius Environmental Bulletin. 2019;**28**(9): 6477-6484